Codes and ciphers

Julius Caesar, the Enigma and the internet

R. F. Churchhouse

PUBLISHED BY THE PRESS SYNDICATE OF THE UNIVERSITY OF CAMBRIDGE
The Pitt Building, Trumpington Street, Cambridge, United Kingdom

CAMBRIDGE UNIVERSITY PRESS
The Edinburgh Building, Cambridge CB2 2RU, UK
40 West 20th Street, New York, NY 10011–4211, USA
477 Williamstown Road, Port Melbourne, VIC 3207, Australia
Ruiz de Alarcón 13, 28014 Madrid, Spain
Dock House, The Waterfront, Cape Town 8001, South Africa

http://www.cambridge.org

First published 2002

Printed in the United Kingdom at the University Press, Cambridge

Typeface Lexicon (*The Enschedé Font Foundry*) 9/13 pt *System* QuarkXPress™ [SE]

A catalogue record for this book is available from the British Library

Library of Congress Cataloguing in Publication data

Churchhouse, R. F.
Codes and ciphers: Julius Caesar, the Enigma, and the Internet / R. F. Churchhouse.
p. cm.
Includes bibliographical references and index.
ISBN 0 521 81054 X – ISBN 0 521 00890 5 (pbk.)
1. Cryptography. 2. Ciphers. I. Title.

Z103 .C48 2002
652′.8–dc21 2001037409

ISBN 0 521 81054 X hardback
ISBN 0 521 00890 5 paperback

Contents

Preface *ix*

1 Introduction 1

Some aspects of secure communication 1

Julius Caesar's cipher 2

Some basic definitions 3

Three stages to decryption: identification, breaking and setting 4

Codes and ciphers 5

Assessing the strength of a cipher system 7

Error detecting and correcting codes 8

Other methods of concealing messages 9

Modular arithmetic 10

Modular addition and subtraction of letters 11

Gender 11

End matter 12

2 From Julius Caesar to simple substitution 13

Julius Caesar ciphers and their solution 13

Simple substitution ciphers 15

How to solve a simple substitution cipher 17

Letter frequencies in languages other than English 24

How many letters are needed to solve a simple substitution cipher? 26

3 Polyalphabetic systems 28

Strengthening Julius Caesar: Vigenère ciphers 28

How to solve a Vigenère cipher 30

Indicators 33

Depths 34

Recognising 'depths' 34

How much text do we need to solve a Vigenère cipher? 37

Jefferson's cylinder 37

4 Jigsaw ciphers 40
Transpositions 40
Simple transposition 40
Double transposition 44
Other forms of transposition 48
Assessment of the security of transposition ciphers 51
Double encipherment in general 52

5 Two-letter ciphers 54
Monograph to digraph 54
MDTM ciphers 56
Digraph to digraph 58
Playfair encipherment 59
Playfair decipherment 60
Cryptanalytic aspects of Playfair 61
Double Playfair 61

6 Codes 64
Characteristics of codes 64
One-part and two-part codes 65
Code plus additive 67

7 Ciphers for spies 72
Stencil ciphers 73
Book ciphers 75
Letter frequencies in book ciphers 79
Solving a book cipher 79
Indicators 86
Disastrous errors in using a book cipher 86
'GARBO''s ciphers 88
One-time pad 92

8 Producing random numbers and letters 94
Random sequences 94
Producing random sequences 95
Coin spinning 95
Throwing dice 96
Lottery type draws 97
Cosmic rays 97
Amplifier noise 97
Pseudo-random sequences 98
Linear recurrences 99
Using a binary stream of key for encipherment 100
Binary linear sequences as key generators 101

Cryptanalysis of a linear recurrence 104
Improving the security of binary keys 104
Pseudo-random number generators 106
The mid-square method 106
Linear congruential generators 107

9 The Enigma cipher machine 110
Historical background 110
The original Enigma 112
Encipherment using wired wheels 116
Encipherment by the Enigma 118
The Enigma plugboard 121
The Achilles heel of the Enigma 121
The indicator 'chains' in the Enigma 125
Aligning the chains 128
Identifying R1 and its setting 128
Doubly enciphered Enigma messages 132
The Abwehr Enigma 132

10 The Hagelin cipher machine 133
Historical background 133
Structure of the Hagelin machine 134
Encipherment on the Hagelin 135
Choosing the cage for the Hagelin 138
The theoretical 'work factor' for the Hagelin 142
Solving the Hagelin from a stretch of key 143
Additional features of the Hagelin machine 147
The slide 147
Identifying the slide in a cipher message 148
Overlapping 148
Solving the Hagelin from cipher texts only 150

11 Beyond the Enigma 153
The SZ42: a pre-electronic machine 153
Description of the SZ42 machine 155
Encipherment on the SZ42 155
Breaking and setting the SZ42 158
Modifications to the SZ42 159

12 Public key cryptography 161
Historical background 161
Security issues 163
Protection of programs and data 163
Encipherment of programs, data and messages 164

The key distribution problem 166
The Diffie–Hellman key exchange system 166
Strength of the Diffie–Hellman system 168

13 Encipherment and the internet 170
Generalisation of simple substitution 170
Factorisation of large integers 171
The standard method of factorisation 172
Fermat's 'Little Theorem' 174
The Fermat–Euler Theorem (as needed in the RSA system) 175
Encipherment and decipherment keys in the RSA system 175
The encipherment and decipherment processes in the RSA system 178
How does the key-owner reply to correspondents? 182
The Data Encryption Standard (DES) 183
Security of the DES 184
Chaining 186
Implementation of the DES 186
Using both RSA and DES 186
A salutary note 187
Beyond the DES 187
Authentication and signature verification 188
Elliptic curve cryptography 189

Appendix 190

Solutions to problems 218

References 230
Name index 235
Subject index 237

Preface

Virtually anyone who can read will have come across codes or ciphers in some form. Even an occasional attempt at solving crosswords, for example, will ensure that the reader is acquainted with anagrams, which are a form of cipher known as *transpositions*. Enciphered messages also appear in children's comics, the personal columns of newspapers and in stories by numerous authors from at least as far back as Conan Doyle and Edgar Allan Poe.

Nowadays large numbers of people have personal computers and use the internet and know that they have to provide a password that is enciphered and checked whenever they send or receive e-mail. In business and commerce, particularly where funds are being transferred electronically, authentication of the contents of messages and validation of the identities of those involved are crucial and encipherment provides the best way of ensuring this and preventing fraud.

It is not surprising then that the subject of codes and ciphers is now much more relevant to everyday life than hitherto. In addition, public interest has been aroused in 'codebreaking', as it is popularly known, by such books and TV programmes as those that have been produced following the declassification of some of the wartime work at Bletchley, particularly on the Enigma machine.

Cipher systems range in sophistication from very elementary to very advanced. The former require no knowledge of mathematics whereas the latter are often based upon ideas and techniques which only graduates in mathematics, computer science or some closely related discipline are likely to have met. Perhaps as a consequence of this, most books on the subject of codes and ciphers have tended either to avoid mathematics entirely or to assume familiarity with the full panoply of mathematical ideas, techniques, symbols and jargon.

It is the author's belief, based upon experience, that there is a middle way and that, without going into all the details, it is possible to convey to non-specialists the essentials of some of the mathematics involved even in the more modern cipher systems. My aim therefore has been to introduce the general reader to a number of codes and ciphers, starting with the ancient and elementary and progressing, via some of the wartime cipher machines, to systems currently in commercial use. Examples of the use, and methods of solution, of various cipher systems are given but in those cases where the solution of a realistically sized message would take many pages the method of solution is shown by scaled-down examples.

In the main body of the text the mathematics, including mathematical notation and phraseology, is kept to a minimum. For those who would like to know more, however, further details and explanations are provided in the mathematical appendix where, in some cases, rather more information than is absolutely necessary is given in the hope of encouraging them to widen their acquaintance with some fascinating and useful areas of mathematics, which have applications in 'codebreaking' and elsewhere.

I am grateful to Cardiff University for permission to reproduce Plates 9.1 to 9.4 inclusive, 10.1 and 10.2, and to my son John for permission to reproduce Plate 11.1. I am also grateful to Dr Chris Higley of Information Services, Cardiff University, for material relating to Chapter 13 and to the staff at CUP, particularly Roger Astley and Peter Jackson, for their helpfulness throughout the preparation of this book.

1

Introduction

Some aspects of secure communication

For at least two thousand years there have been people who wanted to send messages which could only be read by the people for whom they were intended. When a message is sent by hand, carried from the sender to the recipient, whether by a slave, as in ancient Greece or Rome, or by the Post Office today, there is a risk of it going astray. The slave might be captured or the postman might deliver to the wrong address. If the message is written *in clear,* that is, in a natural language without any attempt at concealment, anyone getting hold of it will be able to read it and, if they know the language, understand it.

In more recent times messages might be sent by telegraph, radio, telephone, fax or e-mail but the possibility of them being intercepted is still present and, indeed, has increased enormously since, for example, a radio transmission can be heard by anyone who is within range and tuned to the right frequency whilst an e-mail message might go to a host of unintended recipients if a wrong key on a computer keyboard is pressed or if a 'virus' is lurking in the computer.

It may seem unduly pessimistic but a good rule is to assume that any message which is intended to be confidential *will* fall into the hands of someone who is not supposed to see it and therefore it is prudent to take steps to ensure that they will, at least, have great difficulty in reading it and, preferably, will not be able to read it at all. The extent of the damage caused by unintentional disclosure may depend very much on the time that has elapsed between interception and reading of the message. There are occasions when a delay of a day or even a few hours in reading a message nullifies the damage; for example, a decision by a shareholder to buy or sell a large number of shares at once or, in war, an order by an army commander to attack in a certain direction at dawn next day. On other occasions the information may have long term value and must be kept secret for as long as possible, such as a message which relates to the planning of a large scale military operation.

The effort required by a rival, opponent or enemy to read the message is therefore relevant. If, using the best known techniques and the fastest computers available, the message can't be read by an unauthorised recipient in less time than that for which secrecy or confidentiality is essential then the sender can be reasonably happy. He cannot ever be *entirely* happy since success in reading some earlier messages may enable the opponent to speed up the process of solution of subsequent messages. It is also possible that a technique has been discovered of which he is unaware and consequently his opponent is able to read the message in a much shorter time than he believed possible. Such was the case with the German Enigma machine in the 1939–45 war, as we shall see in Chapter 9.

Julius Caesar's cipher

The problem of ensuring the security of messages was considered by the ancient Greeks and by Julius Caesar among others. The Greeks thought of a bizarre solution: they took a slave and shaved his head and scratched the message on it. When his hair had grown they sent him off to deliver the message. The recipient shaved the slave's head and read the message. This is clearly both a very insecure and an inefficient method. Anyone knowing of this practice who intercepted the slave could also shave his head and read the message. Furthermore it would take weeks to send a message and get a reply by this means.

Julius Caesar had a better idea. He wrote down the message and moved every letter three places forward in the alphabet, so that, in the English alphabet, A would be replaced by D, B by E and so on up to W which would be replaced by Z and then X by A, Y by B and finally Z by C. If he had done this with his famous message

VENI. VIDI. VICI.
(I came. I saw. I conquered.)

and used the 26-letter alphabet used in English-speaking countries (which, of course, he would not) it would have been sent as

YHQL. YLGL. YLFL.

Not a very sophisticated method, particularly since it reveals that the message consists of three words each of four letters, with several letters repeated. It is difficult to overcome such weaknesses in a naïve system like this although extending the alphabet from 26 letters to 29 or more in order to accommodate punctuation symbols and spaces would make the word lengths *slightly* less obvious. Caesar nevertheless earned a place in the history of *cryptography*, for the 'Julius Caesar' cipher, as it is still called, is an early example of an *encryption system* and is a special case of a *simple substitution cipher* as we shall see in Chapter 2.

Some basic definitions

Since we shall be repeatedly using words such as *digraph, cryptography* and *encryption* we define them now.

A *monograph* is a single letter of whatever alphabet we are using. A *digraph* is any pair of *adjacent* letters, thus AT is a digraph. A *trigraph* consists of three adjacent letters, so THE is a trigraph, and so on. A *polygraph* consists of an unspecified number of adjacent letters. A polygraph need not be recognisable as a word in a language but if we are attempting to decipher a message which is expected to be in English and we find the heptagraph MEETING it is much more promising than if we find a heptagraph such as DKRPIGX.

A *symbol* is any character, including letters, digits, and punctuation, whilst a *string* is any adjacent collection of symbols. The *length* of the string is the number of characters that it contains. Thus A3£%$ is a string of length 5.

A *cipher system,* or *cryptographic system,* is any system which can be used to change the text of a message with the aim of making it unintelligible to anyone other than intended recipients.

The process of applying a cipher system to a message is called *encipherment* or *encryption.*

The original text of a message, before it has been enciphered, is referred to as *the plaintext*; after it has been enciphered it is referred to as *the cipher text.*

The reverse process to *encipherment,* recovering the original text of a message from its enciphered version, is called *decipherment* or *decryption.* These two words are not, perhaps, entirely synonymous. The intended recipient of a message would think of himself as *deciphering* it whereas an unintended recipient who is trying to make sense of it would think of himself as *decrypting* it.

Cryptography is the study of the design and use of *cipher systems* including their strengths, weaknesses and vulnerability to various methods of attack. A *cryptographer* is anyone who is involved in *cryptography.*

Cryptanalysis is the study of methods of solving *cipher systems*. A *cryptanalyst* (often popularly referred to as a *codebreaker*) is anyone who is involved in *cryptanalysis.*

Cryptographers and cryptanalysts are adversaries; each tries to outwit the other. Each will try to imagine himself in the other's position and ask himself questions such as 'If I were him what would I do to defeat me?' The two sides, who will probably never meet, are engaged in a fascinating intellectual battle and the stakes may be very high indeed.

Three stages to decryption: identification, breaking and setting

When a cryptanalyst first sees a cipher message his first problem is to discover what type of cipher system has been used. It may have been one that is already known, or it may be new. In either case he has the problem of ***identification***. To do this he would first take into account any available collateral information such as the type of system the sender, if known, has previously used or any new systems which have recently appeared anywhere. Then he would examine the preamble to the message. The preamble may contain information to help the intended recipient, but it may also help the cryptanalyst. Finally he would analyse the message itself. If it is too short it may be impossible to make further progress and he must wait for more messages. If the message is long enough, or if he has already gathered several sufficiently long messages, he would apply a variety of mathematical tests which should certainly tell him whether a code book, or a relatively simple cipher system or something more sophisticated is being used.

Having identified the system the cryptanalyst may be able to estimate how much material (e.g. how many cipher letters) he will need if he is to have a reasonable chance of ***breaking*** it, that is, knowing exactly how messages are enciphered by the system. If the system is a simple one where there are no major changes from one message to the next, such as a codebook, simple substitution or transposition (see Chapters 2 to 6) he may then be able to decrypt the message(s) without too much difficulty. If, as is much more likely, there are parts of the system that are changed from message to message he will first need to determine the parts that don't change. As an example, anticipating Chapter 9, the Enigma machine contained several wheels; inside these wheels were wires; the wirings inside the wheels didn't change but the order in which the wheels were placed in the machine changed daily. Thus, the wirings were the fixed part but their order was variable. The breaking problem is the most difficult part; it could take weeks or months and involve the use of mathematical techniques, exploitation of operator errors or even information provided by spies.

When the fixed parts have all been determined it would be necessary to work out the variable parts, such as starting positions of the Enigma wheels, which changed with each message. This is the ***setting*** problem. When it is solved the messages can be decrypted.

So ***breaking*** refers to the encipherment system in general whilst ***setting*** refers to the decryption of individual messages.

Codes and ciphers

Although the words are often used loosely we shall distinguish between *codes* and *ciphers*. In a *code* common phrases, which may consist of one or more letters, numbers, or words, are replaced by, typically, four or five letters or numbers, called *code groups,* taken from a *code-book*. For particularly common phrases or letters there may be more than one *code group* provided with the intention that the user will vary his choice, to make identification of the common phrases more difficult. For example, in a four-figure code the word 'Monday' might be given three alternative *code groups* such as 1538 or 2951 or 7392. We shall deal with *codes* in Chapter 6.

Codes are a particular type of *cipher system* but not all *cipher systems* are *codes* so we shall use the word *cipher* to refer to methods of *encipherment* which do not use *code-books* but produce the enciphered message from the original plaintext according to some rule (the word *algorithm* is nowadays preferred to 'rule', particularly when computer programs are involved). The distinction between *codes* and *ciphers* can sometimes become a little blurred, particularly for simple systems. The Julius Caesar cipher could be regarded as using a one-page code-book where opposite each letter of the alphabet is printed the letter three positions further on in the alphabet. However, for most of the systems we shall be dealing with the distinction will be clear enough. In particular the Enigma, which is often erroneously referred to as 'the Enigma code', is quite definitely a *cipher machine* and not a *code* at all.

Historically, two basic ideas dominated cryptography until relatively recent times and many cipher systems, including nearly all those considered in the first 11 chapters of this book were based upon one or both of them. The first idea is to shuffle the letters of the alphabet, just as one would shuffle a pack of cards, the aim being to produce what might be regarded as a random ordering, permutation, or anagram of the letters. The second idea is to convert the letters of the message into numbers, taking A = 0, B = 1, ..., Z = 25, and then add some other numbers, which may themselves be letters converted into numbers, known as 'the *key*', to them letter by letter; if the addition produces a number greater than 25 we subtract 26 from it (this is known as (mod 26) *arithmetic*). The resulting numbers are then converted back into letters. If the numbers which have been added are produced by a sufficiently unpredictable process the resultant cipher message may be very difficult, or even impossible, to decrypt unless we are given the key.

Interestingly, the Julius Caesar cipher, humble though it is, can be thought of as being an example of either type. In the first case our 'shuffle' is equivalent to simply moving the last three cards to the front of the pack so that all letters move 'down' three places and X, Y and Z come to the front. In the second case the key is simply the number 3 repeated indefinitely – as 'weak' a key as could be imagined.

Translating a message into another language might be regarded as a form of encryption using a code-book (i.e. dictionary), but that would seem to be stretching the use of the word *code* too far. Translating into another language by looking up each word in a code-book acting as a dictionary is definitely not to be recommended, as anyone who has tried to learn another language knows.* On the other hand use of a little-known language to pass on messages of short term importance might sometimes be reasonable. It is said, for example, that in the Second World War Navajo Indian soldiers were sometimes used by the American Forces in the Pacific to pass on messages by telephone in their own language, on the reasonable assumption that even if the enemy intercepted the telephone calls they would be unlikely to have anyone available who could understand what was being said.

* I recall a boy at school who wrote a French essay about a traveller in the Middle Ages arriving at an inn at night, knocking on the door and being greeted with the response 'What Ho! Without.' This he translated as 'Que Ho! Sans.' The French Master, after a moment of speechlessness, remarked that 'You have obviously looked up the words in the sort of French dictionary they give away with bags of sugar.'

Another form of encryption is the use of some personal shorthand. Such a method has been employed since at least the Middle Ages by people, such as Samuel Pepys, who keep diaries. Given enough entries such codes are not usually difficult to solve. Regular occurrences of symbols, such as those representing the names of the days of the week, will provide good clues to certain polygraphs. A much more profound example is provided by Ventris's decipherment of the ancient Mycenaen script known as Linear B, based upon symbols representing Greek syllables [1.4].

The availability of computers and the practicability of building complex electronic circuits on a silicon chip have transformed both cryptography and cryptanalysis. In consequence, some of the more recent cipher systems are based upon rather advanced mathematical ideas which require substantial computational or electronic facilities and so were impracticable in the pre-computer age. Some of these are described in Chapters 12 and 13.

Assessing the strength of a cipher system

When a new cipher system is proposed it is essential to assess its strength against all known attacks and on the assumption that the cryptanalyst knows what type of cipher system, but not all the details, is being used. The strength can be assessed for three different situations:

(1) that the cryptanalyst has only cipher texts available;
(2) that he has both cipher texts and their original plaintexts;
(3) that he has both cipher and plain for texts *which he himself has chosen.*

The first situation is the 'normal' one; a cipher system that can be solved in a reasonable time in this case should not be used. The second situation can arise, for example, if identical messages are sent both using the new cipher and using an 'old' cipher which the cryptanalyst can read. Such situations, which constitute a serious breach of security, not infrequently occur. The third situation mainly arises when the cryptographer, wishing to assess the strength of his proposed system, challenges colleagues, acting as the enemy, to solve his cipher and allows them to dictate what texts he should encipher. This is a standard procedure in testing new systems. A very interesting problem for the cryptanalyst is how to construct texts which when enciphered will provide him with the maximum information on the details of the system. The format of these messages will depend on how the encipherment is carried out. The second and third situations can also arise if the cryptanalyst has access to a spy in the cryptographer's organisation; this was the case in the 1930s when the Polish cryptanalysts received plaintext and cipher versions of German Enigma messages. A cipher system that cannot be solved even in this third situation is a strong cipher indeed; it is what the cryptographers want and the cryptanalysts fear.

Error detecting and correcting codes

A different class of codes are those which are intended to ensure the *accuracy* of the information which is being transmitted and not to hide its *content*. Such codes are known as *error detecting and correcting codes* and they have been the subject of a great deal of mathematical research. They have been used from the earliest days of computers to protect against errors in the memory or in data stored on magnetic tape. The earliest versions, such as Hamming codes, can detect and correct a *single* error in a 6-bit character. A more recent example is the code which was used for sending data from Mars by the Mariner spacecraft which could correct up to 7 errors in each 32-bit 'word', so allowing for a considerable amount of corruption of the signal on its long journey back to Earth. On a different level, a simple example of an error *detecting*, but not error *correcting*, code is the ISBN (International Standard Book Number). This is composed of either 10 digits, or 9 digits followed by the letter X (which is interpreted as the number 10), and provides a check that the ISBN does not contain an error. The check is carried out as follows: form the sum

> 1 times (the first digit) +2 times (the second digit) +3 times (the third digit) . . . and so on to +10 times (the tenth digit).

The digits are usually printed in four groups separated by hyphens or spaces for convenience. The first group indicates the language area, the second identifies the publisher, the third is the publisher's serial number and the last group is the single digit *check digit*.

The sum (known as the *check sum*) should produce a multiple of 11; if it doesn't there is an error in the ISBN. For example:

1-234-56789-X produces a check sum of

$$1(1)+2(2)+3(3)+4(4)+5(5)+6(6)+7(7)+8(8)+9(9)+10(10)$$

which is

$$1+4+9+16+25+36+49+64+81+100=385=35\times 11$$

and so is valid. On the other hand

0-987-65432-1 produces a check sum of

$$0+18+24+28+30+30+28+24+18+10=210=19\times 11+1$$

and so must contain at least one error.

The ISBN code can *detect* a *single* error but it cannot *correct* it and if there are two or more errors it may indicate that the ISBN is correct, when it isn't.

The subject of error correcting and detecting codes requires some advanced mathematics and will not be considered further in this book. Interested readers should consult books such as [1.1], [1.2], [1.3].

Other methods of concealing messages

There are other methods for concealing the meaning or contents of a message that do not rely on codes or ciphers. The first two are not relevant here but they deserve to be mentioned. Such methods are

(1) the use of secret or 'invisible' ink,
(2) the use of *microdots*, tiny photographs of the message on microfilm, stuck onto the message in a non-obvious place,
(3) 'embedding' the message inside an otherwise innocuous message, the *words* or *letters* of the secret message being scattered, according to some rule, throughout the non-secret message.

The first two of these have been used by spies; the outstandingly successful 'double agent' Juan Pujol, known as GARBO, used both methods from 1942 to 1945 [1.5]. The third method has also been used by spies but may well also have been used by prisoners of war in letters home to pass on information as to where they were or about conditions in the camp; censors would be on the look-out for such attempts. The third method is discussed in Chapter 7.

The examples throughout this book are almost entirely based upon English texts using either the 26-letter alphabet or an extended version of it to allow inclusion of punctuation symbols such as space, full stop and comma. Modification of the examples to include more symbols or numbers or to languages with different alphabets presents no difficulties *in theory*. If, however, the cipher system is being implemented on a physical device it may be impossible to change the alphabet size without redesigning it; this is true of the Enigma and Hagelin machines, as we shall see later. Non-alphabetic languages, such as Japanese, would need to be 'alphabetised' or, perhaps, treated as non-textual material as are photographs, maps, diagrams etc. which can be enciphered by using specially designed systems of the type used in enciphering satellite television programmes or data from space vehicles.

Modular arithmetic

In cryptography and cryptanalysis it is frequently necessary to add two streams of numbers together or to subtract one stream from the other but the form of addition or subtraction used is usually not that of ordinary arithmetic but of what is known as *modular arithmetic*. In modular arithmetic all additions and subtractions (and multiplications too, which we shall require in Chapters 12 and 13) are carried out with respect to a fixed number, known as *the modulus*. Typical values of the modulus in cryptography are 2, 10 and 26. Whichever modulus is being used all the numbers which occur are replaced by their remainders when they are divided by the modulus. If the remainder is negative the modulus is added so that the remainder becomes non-negative. If, for example, the modulus is 26 the only numbers that can occur are 0 to 25. If then we add 17 to 19 the result is 10 since $17+19=36$ and 36 leaves remainder 10 when divided by 26. To denote that modulus 26 is being used we would write

$$17+19\equiv 10 \pmod{26}.$$

If we subtract 19 from 17 the result (-2) is negative so we add 26, giving 24 as the result.

The symbol $\equiv$ is read as 'is congruent to' and so we would say

'36 is congruent to 10 (mod 26)' and '−2 is congruent to 24 (mod 26)'.

When two streams of numbers (mod 26) are added the rules apply to each pair of numbers separately, with no 'carry' to the next pair. Likewise when we subtract one stream from another (mod 26) the rules apply to each pair of digits separately with no 'borrowing' from the next pair.

Example 1.1

Add the stream 15 11 23 06 11 to the stream 17 04 14 19 23 (mod 26).

Solution

	15	11	23	06	11
	17	04	14	19	23
	32	15	37	25	34
(mod 26)	06	15	11	25	08

and so the result is 06 15 11 25 08.

When the modulus is 10 only the numbers 0 to 9 appear and when the modulus is 2 we only see 0 and 1. Arithmetic (mod 2), or *binary arithmetic* as it is usually known, is particularly special since addition and subtraction are identical operations and so always produce the same result viz:

$$\begin{array}{rrrr}
0 & 0 & 1 & 1\\
+0 & 1 & 0 & 1\\ \hline
0 & 1 & 1 & 2\\
\equiv 0 & 1 & 1 & 0
\end{array}
\qquad
\begin{array}{rrrr}
0 & 0 & 1 & 1\\
-0 & 1 & 0 & 1\\ \hline
0 & -1 & 1 & 0\\
0 & 1 & 1 & 0
\end{array} \text{(mod 2) in both cases.}$$

Modular addition and subtraction of letters

It is also frequently necessary to add or subtract streams of letters using 26 as the modulus. To do this we convert every letter into a two-digit number, starting with A = 00 and ending with Z = 25, as shown in Table 1.1. As with numbers each letter pair is treated separately (mod 26) with no 'carry' or 'borrow' to or from the next pair. When the addition or subtraction is complete the resultant numbers are usually converted back into letters.

Table 1.1

A	B	C	D	E	F	G	H	I	J	K	L	M	N	O	P	Q	R	S	T	U	V	W	X	Y	Z
00	01	02	03	04	05	06	07	08	09	10	11	12	13	14	15	16	17	18	19	20	21	22	23	24	25

Example 1.2
(1) Add TODAY to NEVER (mod 26).
(2) Subtract NEVER from TODAY (mod 26).

Solution

(1)
TODAY = 19 14 03 00 24
NEVER = 13 04 21 04 17
Add 32 18 24 04 41 ≡ 06 18 24 04 15 = GSYEP.

(2)
TODAY = 19 14 03 00 24
NEVER = 13 04 21 04 17
Subtract 06 10 –18 –04 07 ≡ 06 10 08 22 07 = GKIWH.

Gender

Cryptographers, cryptanalysts, spies, 'senders' and recipients are referred to throughout in the masculine gender. This does not imply that they are not occasionally women, for indeed some are, but since the majority are men I use masculine pronouns, which may be interpreted as feminine everywhere.

End matter

At the end of the book are the following. First, the mathematical appendix is intended for those readers who would like to know something about the mathematics behind some of the systems, probabilities, analysis or problems mentioned in the text. A familiarity with pure mathematics up to about the standard of the English A-Level is generally sufficient but in a few cases some deeper mathematics would be required to give a full explanation and then I try to give a simplified account and refer the interested reader to a more advanced work. References to the mathematical appendix throughout the book are denoted by M1, M2 etc. Second, there are solutions to problems. Third, there are references; articles or books referred to in Chapter 5 for instance are denoted by [5.1], [5.2] etc.

2

From Julius Caesar to simple substitution

Julius Caesar ciphers and their solution

In the Julius Caesar cipher each letter of the alphabet was moved along 3 places circularly, that is A was replaced by D, B by E ... W by Z, X by A, Y by B and Z by C. Although Julius Caesar moved the letters 3 places he could have chosen to move them any number of places from 1 to 25. There are therefore 25 versions of the Julius Caesar cipher and this indicates how such a cipher can be solved: write down the cipher message and on 25 lines underneath it write the 25 versions obtained by moving each letter 1, 2, 3, ... , 25 places. One of these 25 lines will be the original message.

Example 2.1

The text of a message enciphered by the Julius Caesar System is

VHFX TM HGVX

Decrypt the message.

Solution

We write out the cipher message and the 25 shifted versions, indicating the shift at the left of each line (see Table 2.1), and we see that the cipher used a shift of 19, for the cipher text is shifted 7 places to give the plain and this means that the plaintext has to be shifted $(26-7)=19$ places to give the cipher. It looks very likely, on the assumption that no other shift would have produced an intelligible message, that we have correctly decrypted the message and so there is no point in writing out the remaining lines. This assumption of uniqueness is reasonable when the cipher message is more than five or six characters in length but for *very* short

Table 2.1

Shift	Message
0	VHFX TM HGVX
1	WIGY UN IHWY
2	XJHZ VO JIXZ
3	YKIA WP KJYA
4	ZLJB XQ LKZB
5	AMKC YR MLAC
6	BNLD ZS NMBD
7	COME AT ONCE

messages there is a possibility of more than one solution; for example if the cipher message is

DSP

there are three possible solutions; as shown in Table 2.2. These are not very meaningful as 'messages' although one can envisage occasions when they might convey some important information; for example they could be the names of horses tipped to win races. Primarily, though, they serve to illustrate an important point that often arises: how long must a cipher message be if it is to have a unique solution? The answer depends upon the cipher system and may be anything from 'about four or five letters' (for a Julius Caesar cipher) to 'infinity' (for a one-time-pad system, as we shall see in Chapter 7).

Table 2.2

Shift	Message
2	FUR
8	LAX
15	SHE

Before leaving Julius Caesar here is a rather amusing case of a non-unique solution. In the case of the cipher 'message'

MSG

(which looks like an abbreviation of the word 'message') two possible solutions are shown in Table 2.3, but it is not claimed that the cipher provides a simple way of translating French into English.

Table 2.3

Shift	Message
2	OUI
12	YES

Simple substitution ciphers

In a *simple substitution* cipher the normal alphabet is replaced by a permutation (or 'shuffle') of itself. Each letter of the normal alphabet is replaced, whenever it occurs, by the letter that occupies the same position in the permuted alphabet.

Here is an example of a permuted alphabet with the normal alphabet written above it:

```
A B C D E F G H I J K L M N O P Q R S T U V W X Y Z
Y M I H B A W C X V D N O J K U Q P R T F E L G Z S
```

If this substitution alphabet is used in place of the normal alphabet and we are using a simple substitution cipher then the message

```
COME AT ONCE
```

that we used before would be enciphered to

```
IKOB YT KJIB
```

and an attempt to solve this as a Julius Caesar cipher would be unsuccessful.

Supposing then that a cryptanalyst decided to treat it as a simple substitution cipher would he be able to solve it? He would note that it apparently consisted of three words containing 4, 2 and 4 letters respectively and that the 1st and 9th letters are identical as are the 2nd and 7th letters and the 4th and 10th letters so that, although there are 10 letters in all, there are only 7 *different* letters. It follows that any set of three words in English, or any other language, which satisfy these criteria is a possible solution. Thus the solution might equally well be, among others,

```
GIVE TO INGE
HAVE TO ACHE
```

or

```
SECT IN EAST.
```

None of these look very likely but they *are* valid and show that a short simple substitution cipher may not have a unique solution. This leads us, as already indicated, to an obvious question: 'How many letters of such a cipher does one need to have in order to be able to find a unique solution?' For a simple substitution cipher a minimum of 50 might be sufficient to ensure uniqueness in most cases, but it wouldn't be too easy a task to solve a message of such a short length. Experience indicates that about 200 are needed to make the solution both easy to obtain and unique. We return to this question later.

There are two other points worth noting about the example and the substitution alphabet above. The first point is that the task of decryption was made easier than it need have been because the words in the cipher were separated by spaces, thus giving away the lengths of the words of the original message. There are two standard ways of eliminating this weakness. The first way is to ignore spaces and other punctuation and simply write the message as a string of alphabetic characters. Thus the message above and its encryption become

```
COMEATONCE
IKOBYTKJIB
```

The result of this is that the cryptanalyst doesn't know whether the message contains one word of 10 letters or several words each of fewer letters and, consequently, the number of possible solutions is considerably increased. The disadvantage of this approach is that the recipient of the message has to insert the spaces etc. at what he considers to be the appropriate places, which may sometimes lead to ambiguity. Thus the task of decipherment is made harder for both the cryptanalyst and the recipient.

The second way, which is more commonly used, is to use an infrequent letter such as X in place of 'space'. On the rare occasions when a real X is required it could be replaced by some other combination of letters such as KS. If we do this with the message in the example the message and its encryption become (since X becomes G in the substitution alphabet and X does not occur in the message itself)

```
COMEXATXONCE
IKOBGYTGKJIB
```

The cryptanalyst might now conjecture that G represents a space and so will find the word lengths. In a longer message he would certainly do this, as we shall see shortly. The recipient will now have no ambiguities to worry about but, on the other hand, the task for the cryptanalyst is made easier than in the previous case.

An extension of this idea is to put some extra characters into the alphabet to allow for space and some punctuation symbols such as full stop and comma. If we do this we need to use additional symbols for the cipher alphabet. Any non-alphabetic symbols will do, a typical trio might be $, % and &. It might then happen that, say, in the 29-letter cipher alphabet D gets represented by &, J by $ and S by % whilst 'space', full-stop and comma become, say, H, F and V. Numerals are usually spelled out in full, but, alternatively, the alphabet could be further extended to cope with them if it were desirable. Such extra characters might make the cipher text look more intractable but in practice the security of the cipher would be only slightly increased.

The second point to notice is that two of the letters in the substitution alphabet above, Q and T, are unchanged. Students of cryptography often think at first that this should be avoided, but there is no need to do so if only one or two letters of this type occur. It can be shown mathematically that a random substitution alphabet has about a 63% chance of having at least one letter unchanged in the cipher alphabet (M1). Gamblers have been known to make money because of this, for if two people each shuffle a pack of cards and then compare the cards from the packs one at a time there is a 63% chance that at some point they will each draw the same card before they reach the end of the pack. The gambler who knows this will suggest to his opponent that they play for equal stakes, with the gambler betting that two identical cards *will* be drawn sometime and his opponent betting that they won't. The odds favour the gambler by about 63:37. (It may seem surprising that the chance of an agreement is 63% both for a 26-letter alphabet and for a pack of 52 cards; in fact the chances are not exactly the same in the two cases but they *are* the same to more than 20 places of decimals.)

How to solve a simple substitution cipher

We shall first see how *not* to solve a simple substitution cipher: by trying all the possibilities. Since the letter A in the normal alphabet can be replaced by any of the 26 letters and the letter B by any of the remaining 25 letters and the letter C by any of the remaining 24 letters, and so on, we see that the number of different possible simple substitution alphabets is

$$26 \times 25 \times 24 \times 23 \times \cdots \times 3 \times 2 \times 1$$

which is written in mathematics, for convenience, as 26!, called *factorial* 26. This is an enormous number, bigger than 10 to the 26th power, (or 10^{26} as it is commonly written) so that even a computer capable of testing one thousand million (i.e. 10^9) alphabets every second would take several hundred million years to complete the task. Evidently, the method of trying all possibilities, which works satisfactorily with Julius Caesar ciphers, where there are only 25 of them, is quite impracticable here.

The practical method for solving this type of cipher is as follows.

(1) Make a frequency count of the letters occurring in the cipher, i.e. count how many times A, B, C, ..., X, Y, Z occur.

(2) Attempt to identify which cipher character represents 'space'. This should be easy unless the cipher message is very short, since 'space' and punctuation symbols account for between 15% and 20% of a typical text in English with 'space' itself accounting for most of this. It is highly likely that the most frequently occurring cipher letter represents 'space'. Furthermore, if this assumption is correct, the cipher letter which represents 'space' will appear after every few characters, with no really long gaps.

(3) Having identified 'space', rewrite the text with the spaces replacing the cipher character representing it. The text will now appear as a collection of separated 'words' which are of the same length and structure as the plaintext words. So, for example, if a plaintext word has a repeated letter so will its cipher version.

(4) Attempt to identify the cipher representations of some of the high frequency letters such as E, T, A, I, O and N which will together typically account for over 40% of the entire text, with E being by far the most common letter in most texts.

A table of typical frequencies of letters in English is a great help at this point and such a table is given as Table 2.4; a second table, based upon a much larger sample, will be found in Chapter 7; either will suffice for solving simple substitution ciphers. The tables should only be treated as guides; the higher letter frequencies are reasonably consistent from one sample to another but low letter frequencies are of little value. In the table of English letter frequencies printed below, the letters J, X and Z are shown as having frequencies of 1 in 1000 but in any particular sample of 1000 letters any one of them may occur several times or not at all. Similar remarks apply to letter frequencies in most languages.

(5) With some parts of words identified in this way look for short words with one or two letters still unknown, for example if we know T and E and see a three-letter word with an unknown letter between T and E

then it is probably THE and the unknown letter is H. Recovery of words such as THIS, THAT, THERE and THEN will follow, providing more cipher–plain pairings.

(6) Complete the solution by using grammatical and contextual information.

Table 2.4 *Typical frequencies in a sample of 1000 letters of English text (based on a selection of poems, essays and scientific texts)*

A	57	E	116	I	58	M	14	Q	3	U	25	Y	18
B	9	F	28	J	1	N	57	R	49	V	9	Z	1
C	17	G	14	K	5	O	53	S	55	W	11		
D	26	H	46	L	34	P	18	T	91	X	1		
Punctuation characters													184

Example 2.2

A cipher message consisting of 53 five-letter groups has been intercepted. It is known that the system of encipherment is simple substitution and that spaces in the original were represented by the letter Z, all other punctuation being ignored. Recover the plaintext of the message.

The cipher message is

```
MJZYB LGESE CNCMQ YGXYS PYZDZ PMYGI IRLLC
PAYCK YKGWZ MCWZK YFRCM ZYVCX XZLZP MYXLG
WYTJS MYGPZ YWCAJ MYCWS ACPZY XGLYZ HSWBN
ZYXZT YTGRN VYMJC POYMJ SMYCX YMJZL ZYSLZ
YMTZP MQYMJ LZZYB ZGBNZ YCPYS YLGGW YMJZP
YMJZL ZYCKY SPYZD ZPKYI JSPIZ YMJSM YMJZL
ZYSLZ YMTGY GXYMJ ZWYTC MJYMJ ZYKSW ZYECL
MJVSQ YERMY MJCKY CKYKG
```

Solution

(1) We begin by making a frequency count of the letters: see Table 2.5.

Table 2.5

A	3	E	4	I	4	M	27	Q	3	U	0	Y	49
B	4	F	1	J	17	N	4	R	4	V	3	Z	33
C	18	G	14	K	9	O	1	S	14	W	9		
D	2	H	1	L	14	P	13	T	6	X	8		

(2) Since Y, with 49 occurrences out of 265, is by far the most common letter, accounting for over 18% of the text, we conclude that Y is the cipher representation of the space character. The next most frequent characters are Z and M and we note that these are good candidates for being E and T or T and E.

(3) We now replace Y by 'space' in the cipher text, ignoring the spaces between the five-letter groups, which have no significance, and so obtain a text which reveals the word-lengths. There are 50 words in all in the message and we number them for future reference.

1	2	3	4	5	6	7
MJZ	BLGESECNCMQ	GX	SP	ZDZPM	GIIRLLCPA	CK

8	9	10	11	12	13	14
KGWZMCWZK	FRCMZ	VCXXZLZPM	XLGW	TJSM	GPZ	WCAJM

15	16	17	18	19	20	21	22	23
CWSACPZ	XGL	ZHSWBNZ	XZT	TGRNV	MJCPO	MJSM	CX	MJZLZ

24	25	26	27	28	29	30	31	32	33
SLZ	MTZPMQ	MJLZZ	BZGBNZ	CP	S	LGGW	MJZP	MJZLZ	CK

34	35	36	37	38	39	40	41	42	43
SP	ZDZPK	IJSPIZ	MJSM	MJZLZ	SLZ	MTG	GX	MJZW	TCMJ

44	45	46	47	48	49	50
MJZ	KSWZ	ECLMJVSQ	ERM	MJCK	CK	KG

There are quite a number of short words, the average word length is between 4 and 5 and, overall, the distribution of word lengths looks about right for a natural language, thus lending support to our belief that Y represents the 'space' character.

(4) Looking at the shorter words we find the following.

One word of length 1: word 29, which is S and we guess that S is probably A or I.

Ten words of length 2; one (CK) occurs three times, in positions 7, 33 and 49, and two occur twice – GX in positions 3 and 41, and SP in positions 4 and 34.

Eleven words of length 3, two of which occur twice: MJZ at positions 1 and 44 and SLZ at positions 24 and 39.

(5) Since we already suspect that M and Z are either E and T or vice versa we see that the trigraph MJZ is either E?T or T?E and since it occurs twice

it is very likely that it is THE so that M, J and Z are T, H and E respectively. There are several more words in which the cipher letters M, Z and J are involved including

(23) MJZLZ which becomes THE?E so L is R or S,
(26) MJLZZ which becomes TH?EE which gives L to be R,
(42) MJZW which becomes THE? so W is M or N,
(37) MJSM which becomes THAT if S is A and THIT if S is I.

From these we conclude that L is R and S is A and that W is M or N.

Since word 26 has turned out to be THREE we look at word 25 to see if it could be a number; its cipher form is MTZPMQ which we know is T?E?T? in plain and looks likely to be TWENTY which, if correct, gives T, P, and Q to be W, N and Y respectively and so settles the ambiguity over W which must be M.

(6) We have now identified the plaintext equivalents of nine cipher letters: J, L, M, P, Q, S, W, Y and Z which are H, R, T, N, Y, A, M, 'space' and E. These nine letters together account for over 60% of the text so we would now write out the text again with plaintext equivalents of the cipher letters whenever they are available, otherwise using a dot (.) where the letter is not yet known.

Having done this we would now be able to make some more identifications of cipher–plain pairs. Word 30, which we have partially deciphered as R. . M, has a repeated letter in the middle and can only be ROOM so that cipher letter G is plain letter O. Word 50, KG in cipher is therefore . O in plain which suggests that K represents S, or possibly D, since we already know that it cannot be N or T. Words 48 and 49, MJCK and CK have partially decrypted as TH. S and . S and so lead to the conclusion that C is I. Since C and G occur 18 and 14 times respectively they should be high frequency letters, and I and O are good candidates, as we might have noticed earlier.

Inserting I, O and S for C, G and K in the partially recovered text we have:

```
 1       2         3  4   5      6        7
MJZ BLGESECNCMQ GX SP ZDZPM GIIRLLCPA CK
THE .RO.A.I.IT. O. AN E.ENT O...RRIN. IS

    8         9       10      11   12  13   14
KGWZMCWZK FRCMZ VCXXZLZPM XLGW TJSM GPZ WCAJM
SOMETIMES ..ITE .I..ERENT .ROM .HAT ONE MI.HT
```

```
  15      16     17      18    19      20     21   22   23
CWSACPZ XGL ZHSWBNZ XZT TGRNV MJCPO MJSM CX MJZLZ
IMA.INE .OR E.AM..E .E. WO... THIN. THAT I. THERE

 24    25     26     27    28 29  30   31    32   33
SLZ MTZPMQ MJLZZ BZGBNZ CP S LGGW MJZP MJZLZ CK
ARE TWENTY THREE .EO..E IN A ROOM THEN THERE IS

34  35     36     37    38   39  40  41  42   43
SP ZDZPK IJSPIZ MJSM MJZLZ SLZ MTG GX MJZW TCMJ
AN E.ENS ..AN.E THAT THERE ARE TWO O. THEM .ITH

 44   45     46      47   48   49 50
MJZ KSWZ ECLMJVSQ ERM MJCK CK KG
THE SAME .IRTH.A. ..T THIS IS SO
```

The remaining letters are now easily identified and the entire *decryption* substitution alphabet, denoting 'space' by ^ , is

```
A B C D E F G H I J K L M N O P Q R S T U V W X Y Z
G P I V B Q O X C H S R T L K N Y U A W . D M F ^ E
```

The *encryption* alphabet, which the sender would have used to produce the cipher text from the plain, is of course the inverse of this viz:

```
A B C D E F G H I J K L M N O P Q R S T U V W X Y Z
S E I V Z X A J C . O N W P G B F L K M R D T H Q Y
```

In general the encryption and decryption alphabets will be different in a simple substitution or Julius Caesar system; in the latter case they are the same only when the shift is 13; in the former case they can be made the same by arranging most, if not all, of the letters in pairs so that the letters of a pair encipher to each other, and leaving the remaining letters unchanged. Some cipher machines including both the Enigma and Hagelin machines automatically produce such *reciprocal alphabets*, making the processes of encryption and decryption the same, which is a convenience for the user but also weakens the security. In a simple substitution system based on a 26-letter alphabet the number of possible substitution alphabets is reduced from more than 10^{26} to less than 10^{13}. (For details of this calculation see M2.) Whilst this is still a large number it is significantly less formidable from a cryptanalytic point of view. Such *reciprocal simple substitution ciphers* have, nevertheless, been used occasionally, mainly by individuals who are keeping diaries and wish to make their entries unintelligible to the casual onlooker. The philosopher Ludwig Wittgenstein kept a diary enciphered in this way whilst in the Austrian Army in the 1914–18 war [2.1].

Looking at the example we see that the letter `U` doesn't occur in the cipher text and the letters `J` and `Z` are not present in the plaintext. `Z` was used instead of 'space' in the plaintext and became cipher letter `Y` whilst the letter `J` was the plaintext equivalent of the cipher letter `U` and there is no letter `J` in the original plaintext which is

```
THE PROBABILITY OF AN EVENT OCCURRING IS
SOMETIMES QUITE DIFFERENT FROM WHAT ONE MIGHT
IMAGINE FOR EXAMPLE FEW WOULD THINK THAT IF
THERE ARE TWENTY THREE PEOPLE IN A ROOM THEN
THERE IS AN EVENS CHANCE THAT THERE ARE TWO OF
THEM WITH THE SAME BIRTHDAY BUT THIS IS SO.
```

Those interested in an explanation of this, at first sight remarkable, fact will find it in the mathematical appendix, M3.

Solution of this cryptogram was based partly on the assumption that the frequencies of its individual letters, particularly 'space', `E`, `T`, `A`, `O`, `I` and `N` would be about what one would expect in a sample of such size written in 'typical' English. Sometimes however a passage may be taken from an 'atypical' source, such as a highly specialised scientific work, and so words that one would not find in a novel or newspaper might occur sufficiently often to distort the normal letter frequencies. Studies have been made of millions of characters of English, and other language, texts of different *genres* such as novels, newspaper articles, scientific writing, religious texts, philosophical tracts etc. and the resulting word and letter frequencies published. Brown University in the USA pioneered this work and the tables are given in the 'Brown corpus' [2.2]. Such data are needed for stylistic analysis (trying to determine authorship of anonymous or disputed texts, for example) and other literary studies. A knowledge of the likely subject matter of a cryptogram can be a great help to the cryptanalyst. If he knows, for example, that the message is from one high energy physicist to another words such as `PROTON`, `ELECTRON` or `QUARK` might be in the text and identifying such words in the cipher can substantially reduce the work of decrypting it. Use of unusual words or avoidance of common words can also affect the letter frequencies, which may prove a help or a hindrance to the cryptanalyst. In one extreme case a novel was written which in over 50 000 words never used the letter E, but this was done deliberately; the author having tied down the E on his typewriter so that it couldn't be used. This is a remarkable feat; here, as a sample, is one sentence from the book:

> Upon this basis I am going to show you how a bunch of bright young folks did find a champion; a man with boys and girls of his own; a man of so dominating and happy individuality that Youth is drawn to him as is a fly to a sugar bowl. [2.3]

Even when shown a much longer extract from this book few people notice anything unusual about it until they are asked to study it very carefully and, even then, the majority fail to notice its unique feature.

Letter frequencies in languages other than English

A simple substitution cipher in any alphabetic language is solvable by the method above: a frequency count followed by use of the language itself. Obviously, the cryptanalyst needs to have at least a moderate knowledge of the language, though with a simple substitution cipher he doesn't need to be fluent. Equally obviously the frequency count of letters in a typical sample will vary from one language to another although the variation between languages with a common base, such as Latin, will be less than will be found between languages with entirely different roots. Not all languages use 26 letters; some use fewer; Italian normally uses only 22, and some, such as Russian, use more whilst others (Chinese) don't have an alphabet at all. Since the Italians normally don't use K, W or Y these letters are given a zero frequency, but an Italian text which includes a mention of New York shows that even such letters may appear. In French and German we should really distinguish between vowels with various accents or umlauts but in order to simplify the tables below all forms of the same letter were counted together. Thus, in French, E, É, Ê and È were all included in the count for E. Also, numbers were excluded from the count, unless they were spelled out, and all non-alphabetic symbols such as space, comma, full stop, quotes, semi-colon etc. were considered as 'other'. Upper and lower case letters were treated as the same. With these conventions Table 2.6 shows the frequency of letters in samples of 1,000 in four European languages. The table of frequencies of letters in English given above is repeated for convenience.

Table 2.6

	English	French	German	Italian	Welsh
A	57	72	49	103	77
B	9	13	18	4	13
C	17	17	28	46	23
D	26	34	43	42	63
E	116	143	129	95	55
F	28	7	11	8	28
G	14	11	20	12	32
H	46	9	42	11	43
I	58	56	69	103	57
J	1	1	1	0	0
K	5	0	8	0	0
L	34	42	25	58	47
M	14	35	36	20	23
N	57	54	58	58	58
O	53	48	24	69	64
P	18	27	7	16	3
Q	3	5	0	3	0
R	49	51	69	55	52
S	55	64	54	38	20
T	91	64	64	52	31
U	25	42	28	21	17
V	9	10	8	14	0
W	11	0	12	0	31
X	1	3	0	0	0
Y	18	3	0	0	67
Z	1	1	11	7	0
Other	184	188	186	165	196

A statistical analysis of these counts shows that English, French and German and, to a lesser extent, Italian, are very closely related in so far as single letter frequencies are concerned, whereas their relationship with Welsh is noticeably weaker. A partial explanation is that Y is very common in Welsh, being a vowel (with two different pronunciations), but much less common in English and quite rare in the other languages. The counts also show that N might be said to be 'the most consistent letter' since it occurs with virtually the same frequency in all five languages, accounting for about 6% to 7% of all alphabetic text. For an explanation of the statistical tests typically employed in comparing counts such as these see [2.4]; for further comments see M20.

How many letters are needed to solve a simple substitution cipher?

In Example 2.2 above we had 265 letters available and solved the simple substitution fairly easily. Could we have done so if we had had only, say, 120 letters? More generally, as we have asked earlier, how few letters are likely to be sufficient for a cryptanalyst to solve a cipher such as this? This is a problem in information theory and a formula, which involves the frequencies of the individual letters or polygraphs in the language, has been derived which provides an estimate. The formula is used in an application described in [2.5]. For a simple substitution cipher 200 letters might suffice if we confine our attention to single letters but the use of digraphs (such as `ON`, `IN` or `AT`) or trigraphs (such as `THE` or `AND`) enormously strengthens the attack and it is believed that even 50 or 60 letters might then be enough.

Problem 2.1

An enciphered English text consisting of 202 characters has been found. It is known that a simple substitution cipher has been used and that spaces in the plaintext have been replaced by Z, and all other punctuation ignored. There are reasons for believing that the author preferred to use 'thy' rather than 'your'. Decrypt the text.

```
VHEOC WZIHC BUUCW HDWZB IRWDH TDOZH VIHVI
YBWIU HQOWU HUFWH ZOXBI LHTBI LWDHG DBUWE
HVIRH FVXBI LHGDB UHZOX WEHOI HIODH VCCHU
FPHQB WUPHI ODHGB UHEFV CCHCN DWHBU HSVYJ
HUOHY VIYWC HFVCT HVHCB IWHIO DHVCC HUFPH
UWVDE HGVEF HONUH VHGOD RHOTH BU
```

Example 2.2 illustrates that simple substitution ciphers, though much harder to solve than those of Julius Caesar type, are still too easily solvable to be of much use. For such ciphers the cryptanalyst only requires sufficient cipher text, which corresponds to the first situation mentioned in the previous chapter. Had he been given a corresponding plaintext, as in the second situation, his task would have been really trivial unless the 'message' contained very few distinct letters. In the third situation, where the cryptanalyst is allowed to specify the text to be enciphered, he would simply specify the 'message'

```
ABCDEFGHIJKLMNOPQRSTUVWXYZ
```

and would then have no work to do at all.

To the uninitiated it might seem that since there are more than 10^{26} (i.e. a hundred million million million million) possibilities the task of solving a simple substitution cipher from cipher text alone which, as was pointed out before, would take a computer using the 'brute force' method of trying all of them millions of years, is impossible. We have however just seen how it can be done manually in about an hour by exploiting the known non-random frequencies of the letters and the grammatical rules of English, or whatever is the relevant language, together with any contextual information that might be available. There is a very important lesson in this:

> *it is very dangerous to judge the security of a cipher system purely on the time that it would take the fastest computer imaginable to solve it using a brute force attack.*

The next step, then, is to look at ways of increasing the security of these simple methods and this we do in the next chapter.

3

Polyalphabetic systems

Strengthening Julius Caesar: Vigenère ciphers

The weakness of the Julius Caesar system is that there are only 25 possible decrypts and so the cryptanalyst can try them all. Life can obviously be made more difficult for him if we increase the number of cases that must be tried before success can be assured. We can do this if, instead of shifting each letter by a fixed number of places in the alphabet, we shift the letters by a variable amount depending upon their position in the text. Of course there must be a rule for deciding the amount of the shift in each case otherwise even an authorised recipient won't be able to decrypt the message. A simple rule is to use several fixed shifts in sequence. For example, if instead of a fixed shift of 19 as was used in the message

`COME AT ONCE`

in the last chapter and which enciphered to

`VHFX TM HGVX`

we use two shifts, say 19 and 5, alternately, so that the first, third, fifth etc. letters are shifted 19 places and the second, fourth etc. are shifted 5 places then the cipher now becomes

`VTFJ TY HSVJ.`

If we replace the space character by `Z` in the message and use *three* shifts, say 19, 5 and 11, in sequence the plaintext becomes

`COMEZATZONCE.`

The cipher is now

`VTXXELMEZGHP`

and the key which provides the *encipherment* is 19-5-11. To read the message the recipient must use the *decipherment key* in which each of these three numbers is replaced by its complement (mod 26), i.e. by 7-21-15.

Even if a cryptanalyst suspected that a Julius Caesar system with three shifts being used sequentially was being employed he would have to try 75 or more combinations (one of the shifts might be 0). On such a short message as this there would be the possibility of more than one solution. If the message is too short to identify the three shifts independently, as we shall do in the example below, a 'brute force' method might have to be tried but, since this would involve

$$25 \times 25 \times 25 = 15\,625$$

trials, it would only be used as a last resort. In the extreme case where the number of shifts used was equal to the number of letters in the message the message becomes 'unbreakable', unless there is some non-random feature to the sequence of shifts. Where there is no non-random feature, such as when the sequence of shifts has been generated by some 'random number process', we have what is known as a 'one-time pad' cipher, which we come to in Chapter 7.

This approach to strengthening the Julius Caesar cipher by means of several shifts has been used for some hundreds of years. Such systems are known under the name of Vigenère ciphers. Since most people find it easier to remember words rather than arbitrary strings of letters or numbers Vigenère keys often take the form of a *keyword*. This reduces the number of possible keys of course but that is the price the cryptographer has to pay for easing the burden on his memory. The letters of the keyword are interpreted as numbers in the usual way (A = 0, B = 1, C = 2, ..., Z = 25) so that, for example, the keyword CHAOS would be equivalent to using the five shifts 2, 7, 0, 14 and 18 in sequence.

The keyword or numerical key would be written repeatedly above the plaintext and each plaintext letter moved the appropriate number of places to give the cipher. Thus if we enciphered COMEZATZONCE using Vigenère with the keyword CHAOS the layout would be

```
CHAOSCHAOSCH
COMEZATZONCE
```

and the resultant cipher is

```
EVMSRCAZCFEL.
```

A Vigenère cipher is a particular, and rather special, case of a *polyalphabetic system* in which, as the name implies, a number of different substitution alphabets are used rather than just one, as in simple substitution systems. The number of substitution alphabets used may be anything from 2 to many thousands; in the ENIGMA for example it is effectively 16 900, and these are simple substitutions, not Julius Caesar type shifted alphabets as in Vigenère ciphers, as we shall see in Chapter 9.

How to solve a Vigenère cipher

The first step in solving a Vigenère cipher is to determine the length of the key and, assuming that there is sufficient cipher text available, we do this by looking for repeated combinations of letters, *polygraphs* as they are called, and noting how far apart they are in the text. If these repetitions are genuine, that is if they are cipher versions of the same plaintext, then they will be separated by multiples of the length of the key which should then be identified or, at least, reduced to one of a small number of possibilities. The longer the repeated polygraphs are the better the situation for the cryptanalyst, but even *digraphs*, two-letter combinations, can be helpful.

Example 3.1

A cipher message of 157 characters enciphered by a Vigenère cipher with Z used as 'space' is

```
HQEOT FNMKP ELTEL UEZSI KTFYG STNME GNDGL
PUJCH QWFEX FEEPR PGKZY EHHQV PSRGN YGYSL
EDBRX LWKPE ZMYPU EWLFG LESVR PGJLY QJGNY
GYSLE XVWYP SRGFY KECVF XGFMV ZEGKT LQOZE
LUIKS FYLXK HQWGI LF
```

(1) Find the length of the key.

(2) Find the key and decrypt the message.

Solution

(1) We examine the text and find that six digraphs occur three times or more, viz:

EL at positions 11, 14 and 140;
FY at positions 23, 119 and 146;
GN at positions 31, 64 and 103;
HQ at positions 1, 40, 58 and 151;
LE at positions 70, 91 and 109;
YG at positions 24, 66 and 105.

Further examination reveals that the digraph GN at positions 64 and 103 is in both cases the beginning of an eight-letter ('octograph') repeat:

GNYGYSLE

(these letters have been underlined in the text above).

Eight-letter repeats are very unlikely to occur at random (but Jack Good's experience referred to later in this chapter shows that even 'very unlikely' events do sometimes occur!) so we assume that this is almost certainly significant. We therefore find the distance between the octographs, which is $(103-64)=39$ and since $39=3\times 13$ we conjecture that the key has a length of either 3 or 13. We now look at the distances between repeats of the other digraphs such as the following:

EL at positions 11, 14 and 140 gives intervals of 3 and 126 $(=3\times 42)$;
HQ at positions 1, 40, 58 and 151 gives intervals 39, 18 and 93, all multiples of 3.

These indicate that 3 is by far the most likely length of the keyword.

Assuming that this is so the next step is to find the key.

(2) We now believe that three shifts were used; the first shift being applied to the 1st, 4th, 7th, ... letters; the second shift to the 2nd, 5th, 8th, ... letters and the third shift to the 3rd, 6th, 9th, ... letters. We therefore write the cipher out on a width of three columns and make a frequency count of the cipher letters in each of these three columns and we find Table 3.1. The numbers in the rows total 53, 52 and 52. If the frequencies were randomly distributed each of the numbers should be about 2, but we could reasonably expect a range from 0 to about 5 or 6. Of course the frequencies are far from random since each individual column consists of plaintext letters which have all been shifted by the same amount. The line of attack then is to look for unusually large frequencies in the hope of identifying the letters which are the enciphered versions of Z, the letter used to represent 'space', in the three rows above. In the first row we note that G occurs 13 times and that L, which occurs 7 times, is the next most frequent. If G is the encipherment of Z then the shift for the first row is 7 and L would be the encipherment of E, the letter 7 places before it in the alphabet. Since E is a high frequency letter this lends support to our belief that the first shift is 7, i.e. that the first number in the key is 7.

Table 3.1

Letter	A	B	C	D	E	F	G	H	I	J	K	L	M	N	O	P	Q	R	S	T	U	V	W	X	Y	Z
First shift	0	1	0	0	0	3	13	4	0	0	1	7	1	2	1	5	0	0	2	2	4	4	0	0	2	1
Second shift	0	0	0	0	13	6	0	0	3	2	2	1	2	3	0	0	6	3	3	1	0	0	2	1	3	1
Third shift	0	0	2	2	4	1	1	1	0	1	5	5	1	0	1	4	0	2	3	2	0	1	3	4	6	3

In the second row we see that E occurs 13 times, which makes it a good candidate for being the encipherment of Z, and this would imply that the second shift is 5. In this row the next most frequent cipher letters are F and Q both of which occur 6 times and, shifting these back 5 places, we see that they would correspond to plaintext letters A and L respectively which looks promising. On the other hand the cipher version of plaintext E would be J and this cipher letter only occurs twice in the second row whereas we might expect it to occur 5 times, since E accounts for about 10% of the letters in typical samples of English. The evidence, though not totally convincing, on balance indicates that the second number of the key is probably 5.

In the third row there is no letter of outstandingly high frequency, with Y, K and L which occur 6, 5 and 5 times respectively, being the best contenders as the cipher equivalent of Z. We could try each of these in turn but an alternative approach is to write out the beginning of the cipher text, ignoring the spaces after each five-letter group, and using the assumed shifts of 7 and 5 to decrypt the first and second letters in each group of three. The third letter in each group we 'decrypt' as '/' and we look to see if we can identify any incomplete words and so deduce the third number of the key. So we have

```
Cipher: HQEOTFNMKPELTELUEZSIKTFYGSTNMEGNDGLPUJCHQWFEXFEEPR
Plain:  AL/HO/GH/I /M /N /LD/MA/ N/GH/ I/ G/NE/AL/Y /Y /IM
```

The first word looks as if it is ALTHOUGH and if this is so then plain letter T becomes cipher letter E which implies a shift of 11, since E is 11 places 'on' from T (or, what is the same, 15 places 'behind' T) in the alphabet. With a shift of 11 the cipher letter corresponding to 'space' (i.e. to Z) would be K which was one of our possibilities. We conclude that the third number in the key is 11, so that the three-figure *encipherment* key is 7-5-11, and the *decipherment* key is therefore 19-21-15, and this is confirmed by the full decrypt which is:

ALTHOUGH I AM AN OLD MAN NIGHT IS GENERALLY MY TIME FOR WALKING IN THE SUMMER I OFTEN LEAVE HOME EARLY IN THE MORNING AND ROAM ABOUT FIELDS AND LANES ALL DAY

(which, with punctuation removed, is the opening of Chapter 1 of Dickens's *The Old Curiosity Shop*).

Whilst such polyalphabetic Julius Caesar systems are harder to solve than the simple version they still have the inherent weakness that once we have identified the cipher letter for 'space', or some other high frequency letter, in a shifted alphabet the 25 other letters of that alphabet follow immediately. Then, if we are unsure of the cipher letter for 'space' in one or more alphabets, knowledge of some of the letters in incomplete words may be sufficient to enable us to complete them and so obtain the full solution, as we did in the example above. This weakness disappears if, instead of using alphabets in normal order shifted by different numbers of places according to a key, we use a set of alphabets all in different orders and independent of each other. This modification however raises two other problems:

(1) how do we obtain such alphabets?
(2) how can the shuffled alphabets be made available to the intended recipient(s) without revealing them to unauthorised interceptors?

These questions are of fundamental importance for if the shuffled alphabets are obtained by some simple method, as in Julius Caesar for example, the cryptanalyst will soon spot the method and decryption will be made that much easier. On the other hand the intended recipient must know which alphabets are being used or, alternatively, how to obtain them. There are a number of ways of resolving both of these problems, some of which we shall come to later, but two of the possible solutions to (2) we describe below. First, however, we define two terms that are relevant to cryptographic systems generally.

Indicators

When the originator of a message has a choice of some parameters relating to its encipherment which he needs to make available to the recipient(s) he will probably provide this information, possibly in an enciphered form, in an *indicator* which may precede or follow the message or be hidden within it.

Thus in the example above the key, 7-5-11, would need to be provided somewhere, either in information sent in advance, or enciphered in some pre-arranged cipher and perhaps hidden in the message at some specified place. In this case the key itself, 7-5-11, can serve as the indicator, but it is unlikely to be sent in that form.

Depths

When two or more messages are enciphered by the same system with identical parameters (components, keys, parts, settings etc.) they are said to be *in depth*. So, if two Vigenère messages are sent using the same keyword they are in depth; but if they are sent using different keywords, even if the keywords are of the same length, they are not in depth. If, however, two Vigenère messages have keywords which are of the same length and which have some identical letters in corresponding positions the cipher messages will be in *partial depth*. This would not necessarily be true of other systems of encipherment where the slightest change in the indicator puts the messages out of depth. Whether a cryptanalyst can take advantage of finding two or more messages in depth depends upon the system which has been used for enciphering them. In some cases, such as simple substitution or Vigenère systems, he should certainly be able to do so but in others, such as the two-letter cipher systems described in Chapter 5, depths are of much less use. Broadly speaking, if the encipherment system is done on a letter-by-letter basis then depths may be identifiable and useful to the cryptanalyst but if the encipherment involves two or more letters at a time the depths, if recognisable at all, may not be of much use.

Recognising 'depths'

How would a cryptanalyst recognise a depth? If two or more messages were sent on the same system and were found to have the same indicators they are probably in depth. We must say 'probably', not 'certainly', because the interval between the two transmission times may have overlapped a change-over time when some part of the enciphering system may have been changed. Such a situation would occur, for example, with two Enigma messages sent just before and just after midnight (see Chapter 9).

If the indicators are hidden there may be no external evidence that the messages *are* in depth. So how might the cryptanalyst discover that they are? Assuming that the encipherment system involves only one letter at a time he would first write the messages under one another, lining up the beginning cipher letters of the messages, and then apply a simple statistical test. If two cipher messages are *not* in depth the probability that a cipher letter of one message will be the same as the corresponding cipher letter in the other, i.e. underneath it, is 1 in 26. If the messages *are* in depth the probability that corresponding cipher letters will be the same is equal to the probability that the corresponding *plaintext* letters are the same and this is found to be about twice the random probability, that is, about 1 in 13 for English and for most languages using the Latin alphabet. This is a particular case of a more general observation that we shall come to in Chapter 7; for the mathematical details see M6. It follows that if we have aligned a pair of messages we would expect about four cases of identical cipher pairs per hundred cipher characters if the messages *are not* in depth but about seven or eight identical pairs per hundred characters if the messages *are* in depth. The longer the messages the stronger the evidence for or against a depth. The case for a depth is strengthened considerably if there are any polygraph coincidences, e.g. 2- or 3-letter agreements, since these are very unlikely in messages which are not in depth. Of course this test is not infallible, polygraph coincidences *can* occur at random in pairs of cipher texts. Jack Good in [3.1] records that he once found a completely bogus *octagraph* in a pair of wartime messages. The probability of an 8-long coincidence such as this is less than 1 in 20 000 000 000. Even allowing for all the cipher messages looked at during the War this seems remarkable. On the other hand Jack also records [3.1] that he also found a 22-long repeat; this *was* genuine!

Deciding whether a pair of messages are in depth gets easier as the length of the shorter message increases. Thus, it is easier to identify a depth with a pair of messages each of which is 500 letters in length than in a pair of messages one of which has 2000 letters whilst the other is only 100 letters long. It is the length of the *overlap* between the messages that matters.

Example 3.2

Three messages have been enciphered using a system that enciphers one letter at a time. The shortest message is 500 characters in length and the numbers of cipher letter agreements between the three pairs of messages are found to be as follows.

Message 1 and message 2: 37.
Message 1 and message 3: 27.
Message 2 and message 3: 16.

Are any pair of these messages likely to be in depth?

Solution

With a 500-letter overlap we would expect about 38 cipher letter agreements if a pair of messages are in depth and only about 19 if they are not in depth. On this basis the data surely imply that messages 1 and 2 are in depth and that messages 2 and 3 are not in depth. The evidence for the pairing of messages 1 and 3 is anomalous and on the face of it is inconclusive since the probability of seeing 27 agreements when we expect 38 is about the same as the probability of seeing 27 when we expect 19 (the mathematical basis for this can be found in books such as [2.4]). However, since we are fairly sure that messages 1 and 2 *are* in depth and that messages 2 and 3 *are not* we can reasonably conclude that messages 1 and 3 are not.

From a cryptanalyst's point of view the most valuable depths would be those where the encipherment system involved adding a stream of key to the plaintext such as in book ciphers or one-time pad (Chapter 7) or, at a simpler level, Vigenère. The otherwise 'unbreakable' one-time pad becomes breakable when a depth is found.

Messages sent in Vigenère systems with *different* keys can sometimes produce unusual features which will help the cryptanalyst to identify them as, can be seen from the following.

Problem 3.1

A music-loving spymaster sent three of his agents an identical plaintext message using Vigenère encipherment with the keywords

(1) `RHAPSODY`, (2) `SYMPHONY` and (3) `SCHUBERT`.

What should a cryptanalyst find in examining the pairs of cipher text?

Check your conclusions by enciphering (using `X` as the separator) the message

```
NOW IS THE TIME FOR ALL GOOD MEN TO COME TO
THE AID OF THE PARTY
```

using the three keywords and comparing the three cipher messages in pairs.

Before leaving Vigenère try the following.

Problem 3.2

A message of 249 letters with Z used as space has been enciphered using a Vigenère cipher. The text is

```
GLEKR DAKRD SHZIZ MUIOK RQSSJ MTAME ZIESO
YMAHB PLZBF DSHMW HHEXZ TAHZX YIGTA XZMUE
TSVXZ LRIML MYNEV OEELD TANXZ TMFEM GIRSB
RESJM LEMIV XEDBX MJONA HZLHG HSVWZ MUIZV
NWESJ MTAMI UVYMD LMTRH BJZMU ETSGL EKRDA
KRDAG MMNYV RIMRD NNZFE KMSFS CVIFR WZMUM
SSCVO HSDIL MMNSG LESNT PXAHI QMMNS GLILM
FOHX.
```

Find the key and decrypt the message.

How much text do we need to solve a Vigenère cipher?

In Example 3.1 we had 157 characters of cipher and a key of length 3 so that we had over 50 characters from each cipher alphabet. With this much text we found numerous repeated digraphs, some of which extended into trigraphs and, rather luckily, one octograph. From these we were able to find the key and decipher the message quite easily. A cipher text of 50 times the length of the key should, in general, be adequate to solve a cipher of this type. The Vigenère system is therefore vulnerable under situation (1) of Chapter 1. In situation (2), where the plain and cipher texts are both available a text of length twice the length of the key would be sufficient. Obviously, Vigenère ciphers cannot be recommended unless either the messages are very short or the keys are very long.

Jefferson's cylinder

A simple device which provides a series of simple substitution alphabets seems to have been constructed by Thomas Jefferson in the late eighteenth century and subsequently re-invented by others. The device is made of a set of numbered, physically identical, discs mounted on a common axis about which they can be rotated independently. Each disc has the alphabet in some shuffled order, probably different for each disc, engraved on its periphery.

There could, in theory, be any number of discs in the set but there would typically be between 20 and 40. If, for example, there were 30 discs in the set the message to be enciphered would first be broken up into blocks each containing 30 letters. The 30 discs on the cylinder would then be rotated so that the first 30 letters of the message appeared in a horizontal line and then the 30 letters of *any* of the other 25 horizontal lines would be sent as the cipher. The discs would then be rotated so that the second block of 30 plaintext letters lined up and, again, any one of the other 25 horizontal lines chosen as the cipher text. Decipherment would involve the recipient lining up the discs to show the 30 letters of each block of cipher in a horizontal line and then looking at the other 25 horizontal lines and finding the one that made sense. This raises the question: 'Is it possible that there might be more than one line that makes sense?' Such a situation cannot arise with most cipher systems, where there is a unique relationship between the plaintext and cipher text, but the Jefferson cylinder is an exception. If the plaintext message is in a natural language the possibility of finding more than one sensible line among the 25 is negligible. On the other hand if the plaintext message consists of code groups, such as are described in Chapter 6, it is not impossible that more than one valid decipherment could be found. It would depend upon how many of the theoretically available code groups were actually used in the code. If *all* possible code groups were being used every one of the 25 lines would produce a valid decrypt, though it is unlikely that more than one of them would make sense when converted back to natural language.

The security of the Jefferson cylinder would be considerably enhanced if the order in which the discs were placed on the common axis could be changed regularly, e.g. daily, but this would necessitate the sender(s) and recipient(s) agreeing on the ordering in advance. In this case the ordering of the discs would be 'the indicator' for the messages but, instead of being given in the preamble or embedded in the text of the message, it would probably be given in a printed sheet which had been produced and distributed some time before the start of the cipher-period.

Without knowing the shuffled alphabets on the discs or the order in which the discs themselves were placed on the common axis the cryptanalyst would have considerable difficulty in decrypting messages and would probably need many messages sent in the same cipher-period (i.e. using the same ordering of the discs) to achieve success. If the order of the discs cannot be changed the cipher-period is infinite and all messages sent using this system can be considered together but decryption will still not be easy since, although the messages can all be lined up vertically in blocks of 30 letters, they are not strictly *in depth* because the same letter in the same position in a block may be enciphered to any one of 25 letters. Once the shuffled alphabets have been recovered, however, possibly through senders making mistakes or some plaintext messages being obtained, decryption becomes trivial.

Simple in concept though it is, the Jefferson cylinder provided a genuinely polyalphabetic cipher system and was an ingenious and effective form of cipher machine. As such, it was a forerunner of some of the most widely used cipher machines of the twentieth century, as we shall see later.

4

Jigsaw ciphers

In this chapter we look at a number of cipher systems which are based upon a different idea to those that we have met so far. In these systems each letter retains its own identity and so the frequencies of the individual letters of the messages are unchanged but the constituent letters of the *digraphs*, and the higher order polygraphs, are separated and, consequently, their original plaintext frequencies are destroyed. Since the method used in trying to solve them is rather like that of piecing together a jigsaw I have grouped them under the (unofficial) name of 'jigsaw ciphers'. The simplest such systems are called

Transpositions

The cipher systems that we have examined in the earlier chapters have been based on substitution alphabets, where each letter is replaced by another letter but the order of the letters in a message is unchanged. An alternative approach is to leave the letters of the message unaltered but change their order. The result is that the cipher message is an anagram of the plaintext message. The simplest way of doing this to use a *transposition system*.

Simple transposition

In a *simple transposition system* the message is first written into a box, usually a rectangle, which has been divided up into squares by a number of horizontal and vertical lines. The number of *vertical* lines is fixed by a numerical or literal key; the number of *horizontal* lines may be fixed or may be determined by the length of the message. If the number of rows is fixed the message is broken up into stretches of the appropriate length, the capacity of the box. The message is written into the box *row by row*, beginning at the top, but it is then taken out of the box *column by column* in an order determined by the key. The result is that the letters are unchanged but they are transmitted in a different order. The method of encipherment is very simple as the following example shows.

Example 4.1

Use a simple transposition system with the 5-digit key 3-1-5-2-4 to encipher the message

MEETING WILL BE ON FRIDAY AT ELEVEN THIRTY

Encipherment

Ignoring the spaces between the words, there are 35 letters in the message. Since the key is of length 5 we will need a box with 5 columns and 7 rows.

Table 4.1

Key	3	1	5	2	4
	M	E	E	T	I
	N	G	W	I	L
	L	B	E	O	N
	F	R	I	D	A
	Y	A	T	E	L
	E	V	E	N	T
	H	I	R	T	Y

We therefore write down the key and underneath it we draw a box with 5 columns and 7 rows, to hold the message. We then write the message into the box row by row, beginning in the top row, and ignoring spaces between words: see Table 4.1. Finally, we write out the message as a series of 5-letter groups, taking the text out of the box column by column in the order indicated by the key:

EGBRA VITIO DENTM NLFYE HILNA LTYEW EITER.

This is what both the recipient and cryptanalyst will see. How will the recipient decipher the message, and how might a cryptanalyst set about trying to solve it?

Decipherment

The recipient writes the cipher message into the transposition rectangle column by column in the column order given by the transposition key and then reads the message row by row, beginning at the top.

Cryptanalytic attack

Simple though it is, a transposition cipher may not be easy to solve. A frequency count of the individual letters ('monographs') will reveal that they have not been changed but the frequencies of pairs of letters ('digraphs') such as TH, HE and QU will be different from what would be expected in an English text. The cryptananlyst would therefore suspect that a transposition system is being used and his first task in trying to solve it would be to determine the length of the key.

Since the message above is 35 letters in length all 7 groups have 5 letters. The cryptanalyst would not know whether the message was genuinely 35 letters in length or had been 'padded out' with some 'dummy letters' in order to produce full cipher groups, all of which have 5 letters. In either case however he has a possible clue as to the length of the key. Since $35 = 5 \times 7$ it is worth looking at the cipher text on the assumption that the key is of length 5 or 7. He is making the assumption that the transposition box is 'regular', i.e. that all the columns are of equal length; they may not be, but this is a reasonable first step.

Assuming that the key *is* of length 5, two letters which were adjacent in the original message will be 7, 14, 21 or 28 positions apart in the cipher text unless one of the pair was at the end of one row and the other was at the beginning of the next. The cryptanalyst would therefore write out the cipher text in 5 columns of 7 letters, which has the effect of giving letters which were next to each other in the plaintext a good chance of being in the same row of the cipher text arranged in this way. Using this arrangement the cipher text becomes as shown in Table 4.2.

Table 4.2

E	T	M	I	E
G	I	N	L	W
B	O	L	N	E
R	D	F	A	I
A	E	Y	L	T
V	N	E	T	E
I	T	H	Y	R

The next step is to look at the various pairs of letters in each row to see which of them look to be the most likely digraphs. For this purpose access to a frequency count of digraphs in English, such as can be found in the Brown corpus and in some books on cryptography, is a great help. In the first row above, for example, we find that the pairs ME, TI and ET are more common than the others. We do this for each row and note the implications; thus if, in the first row, T should really be next to I then column 2 should be to the left of column 4. By carrying out this exercise for each row we hope to find confirmations of such possibilities and so recover the transposition rectangle and the original text. Not every implication will be correct, and some will contradict others, but, hopefully, enough correct implications will be seen to enable us to override the incorrect ones. The evidence is as shown in Table 4.3.

Table 4.3

Row	Digraphs	Implied adjacent columns
1	ME, TI, ET	3-1 or 3-5, 2-4, 1-2 or 5-2
2	IN, WI, NG	2-3, 5-2, 3-1
3	BE, ON, LE	1-5, 2-4, 3-5
4	RI, DA, ID	1-5, 2-4, 5-2
5	AT, ET, EA	1-5, 2-5, 2-1
6	VE, NT	1-3 or 1-5, 2-4
7	IT, TH, HI	1-2, 2-3, 3-1

Although there are some contradictions certain implications occur sufficiently often to merit further examination, namely

1-5, 2-4, 3-1, and 5-2.

If we now rearrange the cipher columns in the order suggested by these pairings we see, when we begin with column 3 (we would have to try other columns in the first position), that the cipher text should be rearranged so that cipher columns

3-1-5-2-4

become plaintext columns

1-2-3-4-5

so that the first cipher row

E T M I E

becomes plaintext row

M E E T I.

Likewise the second cipher row

```
G I N L W
```

becomes

```
N G W I L
```

and the rest of the message confirms the decryption.

This is a very simple example in that the key was short and its length was the first obvious one to try, but it illustrates the method of solution. It also indicates that access to a table of digraph frequencies, though not strictly essential, will make the task much easier. Had the cipher text not been a multiple of 5 it would not have been so likely that the key length might be 5 (or 7, in this case) and other key lengths might have to be tried. For keys of length no longer than 5 even a brute force attack is feasible since there are only a moderate number of possible orderings of the columns (120 when the key length is 5). As key lengths increase beyond 5 a brute force attack soon becomes very tedious and, eventually, impractical by hand, whereas the digraph method used above is realistic for all key lengths that are likely to be encountered in practice. The cryptographer, knowing all this, would therefore attempt to disguise the key length as far as possible and might also resort to other measures, such as

Double transposition

The weakness of the simple transposition method is that when the cipher message is written out column by column 'on the width of the key', i.e. in rows containing as many letters as the key length, letters which were adjacent in the plaintext will tend to fall in the same row and a search for 'good' digraphs may reveal the transposition order of the columns.

This becomes very obvious if we replace the plaintext in the example above by the numbers 1, 2, 3, ..., 35, underlining the first five numbers for ease of identification, and apply the transposition key that we used before, thus (see Table 4.4) we get the 'cipher' text

$\underline{2}$ 7 12 17 22   27 32 $\underline{4}$ 9 14   19 24 29 34 $\underline{1}$ 6 11 16 21 26
31 $\underline{5}$ 10 15 20   25 30 35 $\underline{3}$ 8   13 18 23 28 33

Table 4.4

Key	3	1	5	2	4
	1	2	3	4	5
	6	7	8	9	10
	11	12	13	14	15
	16	17	18	19	20
	21	22	23	24	25
	26	27	28	29	30
	31	32	33	34	35

and if we look at the intervals between the underlined numbers we find that

1 and 2 are 14 places apart
2 and 3 are 28 places apart
3 and 4 are 21 places apart

and

4 and 5 are 14 places apart

i.e. the underlined numbers fall at intervals of multiples of 7, so that when we arrange the 'text' in 7 rows of 5 they fall into the same row.

If however we take this 'cipher' text and apply the transposition again (Table 4.5)

Table 4.5

Key	3	1	5	2	4
	2	7	12	17	22
	27	32	4	9	14
	19	24	29	34	1
	6	11	16	21	26
	31	5	10	15	20
	25	30	35	3	8
	13	18	23	28	33

the 'cipher' text is

7 32 24 11 5 30 18 17 9 34 21 15 3 28 2 27 19 6 31 25
13 22 14 1 26 20 8 33 12 4 29 16 10 35 23

and we see that pairs which were originally adjacent in the 'plaintext' are now non-uniformly distributed in the 'cipher' text,

1 and 2 are 9 places apart
2 and 3 are 2 places apart
3 and 4 are 17 places apart

and

4 and 5 are 25 places apart

and the digraph attack that was used before will no longer work.

The security of the double transposition cipher would be further increased if we used two *different* keywords rather than using the same one twice, particularly if the keywords were of different lengths. In such a system, however, there is the increased risk that the sender will make a mistake by using the two keywords in the wrong order. This will, in general, produce a different cipher text which the receiver won't be able to decipher and the message will have to be sent again in its correct form. This would provide the cryptanalyst with two versions of the same text, enciphered with the same keys, but in opposite orders; a situation he may be able to exploit. This type of error is a hazard of any system involving double encryption. Some Enigma messages were enciphered twice, on different settings, for extra security and there was at least one occasion when the encipherments were carried out in the wrong order [4.1].

Example 4.2
Apply the double transposition method to the text

A B C D E F G H I J K L M N O

using the two keys

3-1-5-2-4 and 3-1-2

(i) in the order given, (ii) in the opposite order. Hence verify that the two cipher texts are different.

Verification
(i) We apply 3-1-5-2-4 (see Table 4.6),

Table 4.6

3	1	5	2	4
A	B	C	D	E
F	G	H	I	J
K	L	M	N	O

giving

B G L D I N A F K E J O C H M.

Now we apply 3-1-2 (see Table 4.7),

Table 4.7

3	1	2
B	G	L
D	I	N
A	F	K
E	J	O
C	H	M

and the cipher text is

G I F J H L N K O M B D A E C.

(ii) we apply 3-1-2 (see Table 4.8),

Table 4.8

3	1	2
A	B	C
D	E	F
G	H	I
J	K	L
M	N	O

giving

B E H K N C F I L O A D G J M.

Now we apply 3-1-5-2-4 (see Table 4.9),

Table 4.9

3	1	5	2	4
B	E	H	K	N
C	F	I	L	O
A	D	G	J	M

and the cipher text is

E F D K L J B C A N O M H I G

which is quite different from the first version.

Other forms of transposition

So far we have written the text in rectangles row by row but there are alternatives available and some of them would make it harder for a cryptanalyst to solve the system. In each case the text of the message is broken up into blocks of the appropriate length to fit the transposition box.

Regular transposition boxes

By a *regular transposition box* we mean a box that has columns of predictable length. The text can be written into the box in various ways but care needs to be taken, otherwise weaknesses may be introduced. Here are some examples, not all of which can be recommended.

Table 4.10

A	B	C	D	E
P	Q	R	S	F
O	X	Y	T	G
N	W	V	U	H
M	L	K	J	I

(i) We could use a rectangle but write the text around the rectangle in a 'spiral' fashion (Table 4.10). Unfortunately this is definitely not recommended. No matter what ordering of the columns is used in the transposition the pentagraph EFGHI and the trigraph STU will appear in unchanged form in the 'cipher'. So, for example, if the message is

THISXMETHODXISXNOTXSECURE

and the transposition is 3-1-5-2-4 the box is as shown in Table 4.11,

Table 4.11

3	1	5	2	4
T	H	I	S	X
N	O	T	X	M
X	R	E	S	E
S	U	C	E	T
I	X	D	O	H

then the cipher text is

HORUX SXSEO TNXSI XMETH ITECD

The pentagraph, XMETH, would soon be noticed by the cryptanalyst. The trigraph, XSE, though genuine, is not so obvious. If a rectangle with more rows than columns were used the 'good' polygraphs would be even more obvious.

(ii) Use a 'diamond-shaped' box. In this box all the rows (and columns) contain an *odd* number of letters, starting with 1 and increasing each time by 2 up to some pre-determined number and then decreasing by 2 each time to 1. The box is obviously symmetric about the central row and column and all the columns are correctly aligned vertically (Table 4.12).

Table 4.12

			A			
		B	C	D		
	E	F	G	H	I	
J	K	L	M	N	O	P
	Q	R	S	T	U	
		V	W	X		
			Y			

This has the advantage that the column lengths are variable, which complicates the digraph attack. If we use the 7-digit transposition key 3-1-7-5-2-4-6 for example (Table 4.13)

Table 4.13

3	1	7	5	2	4	6
			A			
		B	C	D		
	E	F	G	H	I	
J	K	L	M	N	O	P
	Q	R	S	T	U	
		V	W	X		
			Y			

the transmitted text will be

EKQDH NTXJI OUACG MSWYP BFLRV

and letters which were originally adjacent will now be at distances ranging from 5 (e.g. H–I) to 21 (E–F) from each other, rather than at multiples of 5 as with a normal 5×5 square.

(iii) Use a hexagon. In the example in Table 4.14 each row or column contains an *even* number of letters starting with 2 and increasing by 2 each time, up to some limit. The length of the longest row is repeated and then the lengths decrease steadily to 2. Other designs may be used.

Table 4.14

		A	B		
	C	D	E	F	
G	H	I	J	K	L
M	N	O	P	Q	R
	S	T	U	V	
		W	X		

This offers the same advantage as the diamond box. In this case the transmitted text when the key is 2-5-4-1-6-3 is

BEJPU XGMLR ADIOT WCHNS FKQV.

Irregular transposition boxes

There are many other possibilities, including 'incomplete' rectangles with columns of different lengths, all of which add to the security of the transposition system. In the case of Example 4.1 a transposition box of 35 cells need not consist of 5 columns of 7 letters, or of 7 columns of 5 letters, but might have 5 columns of different lengths, such as 10, 4, 11, 7, and 3, so that the 35-letter message of the example would be entered as shown in Table 4.15, in which case the cipher message would be

EGBRT IODAL EMNLF ATENH RILNE WEIYE VTITY.

Table 4.15

3	1	5	2	4
M	E	E	T	I
N	G	W	I	L
L	B	E	O	N
F	R	I	D	–
A	–	Y	A	–
T	–	E	L	–
E	–	V	E	–
N	–	T	–	–
H	–	I	–	–
R	–	T	–	–
–	–	Y	–	–

The plaintext digraphs are now separated at a variety of distances and since there are 46376 ways of choosing 5 positive numbers which add up to 35 (see the solution to Problem 4.2 below) the cryptanalyst would be faced with a formidable problem. Depths would not be of much use, but if the same message was sent to two recipients using the same box, but different transposition keys, the system might be solved. In any event the recipient, of course, would have to know both the column lengths and the transposition key and if these were suspected of being hidden in an indicator the cryptanalyst would certainly endeavour to find it.

In choosing column lengths for an irregular transposition box it would be wise to ensure that the longest column was not more than, say, two letters longer than the second longest column for otherwise the last few letters of the message will all come in sequence in the cipher text. So, for example, the last two letters in the irregular box above (TY) came together at the end of the cipher text. That they fell at the *end* was a consequence of the particular transposition but they would have come together *somewhere* since the longest column (11) came after the second longest column (10) in the transposition. If the longest column was much longer than the second longest the cryptanalyst would notice a good (high frequency) plaintext polygraph in the text which would give him an important clue.

Assessment of the security of transposition ciphers

A double transposition system with different transpositions, or a simple transposition system with an irregular-shaped box, would not be easily broken unless several cipher messages were available whereas a simple system with a regular box should be solvable, given a sufficiently long message, by the digraph attack. In the situation where both plain- and cipher texts are available even the double system should be solvable if the message(s) contain(s) some letters of low frequency so that they could be uniquely identified in the plain- and cipher texts but, as Example 4.2 indicates, it would not be a trivial task. In the third situation, where the cryptanalyst is allowed to specify the plain texts, he would ensure that no character would appear more than once if possible, which would enable him to see what transformations of the columns would produce the observed shifts of the letters. In all cases the cryptanalyst's first task would be to find the lengths of the transposition keys; until this is done he cannot solve the cipher. For infrequent use in low security messages a double transposition system where the transpositions are changed regularly would be moderately secure. For heavy use or high security messages such systems are not acceptable.

Double encipherment in general

We have seen that the use of double transposition will normally increase the security of a transposition system but at the risk of the transpositions being applied in the wrong order. This raises the general question as to whether enciphering a message twice or more will increase its security. There is no simple answer to this question since it depends upon the type of encipherments employed. In the case of double simple substitution for example there is no advantage because the result of the double substitution is just another simple substitution and so offers no greater challenge to the cryptanalyst. Furthermore if the two substitutions are applied in the wrong order a different cipher text is produced; this won't defeat the cryptanalyst, who won't even be aware of the error, but the genuine recipient won't be able to decipher the message. A combination of simple substitution and transposition on the other hand does offer increased security. A cryptographer who is thinking of using a multiple encipherment system would be sure to ask himself:

(1) does it increase the security?
(2) if the encipherments are being carried out manually would the users find the system to be excessively tedious?
(3) if the encipherments are used in the wrong order would it help the cryptanalyst?

An interesting example of a *triple* encipherment cipher is one of those used by the double agent GARBO. This is described in Chapter 7.

Problem 4.1
The 30-long message below, enciphered by simple transposition, has been sent by a young man to his girl-friend. The key length is believed to be 6; all spacing and punctuation have been ignored. Find the key and decrypt the message.

```
LPEUD SCEOE LAEMA AMHSS HOTAR IRTMY
```

Problem 4.2
The number of possible (regular and irregular) transposition boxes of given width grows very rapidly as the capacity (number of letters) of the

box increases. By enumerating the possibilities show that there are 28 when the width is 3 and the capacity is 9.

Problem 4.3

The plaintext in a transposition system could be written into a square, or rectangular, box along the rows alternately from left to right and then right to left (this is known as *boustrophedon* fashion since it is the path taken in ploughing). Entering the message

THISXMETHODXISXNOTXSECURE

in this way gives Table 4.16.

Table 4.16

THISX
OHTEM
DXISX
SXTON
ECURE

What serious weakness does this system introduce?

5

Two-letter ciphers

It may seem natural that a cipher, unlike a code, will have the property that each individual letter is separately enciphered and replaced by a single letter, or symbol, and this is indeed the case for many cipher systems. There are, however, encryption systems in which each plaintext letter becomes more than one letter in the cipher and there are also other systems which encipher the text two (or more) letters at a time. An example of the first type is

Monograph to digraph

The alphabet is written into a 5×5 square with one letter being omitted The omitted letter is usually J, which is replaced by I if required. The five rows of the square are identified by the letters A, B, C, D and E as also are the columns: see Table 5.1.

Table 5.1

	A	B	C	D	E
A	A	B	C	D	E
B	F	G	H	I	K
C	L	M	N	O	P
D	Q	R	S	T	U
E	V	W	X	Y	Z

Each letter in the plaintext is now replaced by its row and column letters so that, for example, M becomes CB. The resulting cipher text is therefore twice as long as the original plain and contains only the letters A to E.

Example 5.1

HAPPY BIRTHDAY

enciphers as

BCAAC ECEED ABBDD BDDBC ADAAE D

As it stands this is a weak cipher but its security can be improved by two modifications

(i) using a shuffled alphabet inside the 5×5 box;
(ii) applying a transposition to the enciphered text.

The first of these is sometimes partially achieved by choosing a keyword, writing this into the box and then filling the rest of the box with the unused letters of the alphabet, excluding J. If the keyword has any repeated letters they are ignored.

Example 5.2
Repeat the example above using a box with the keyword THURSDAY and apply the transposition 5-1-4-2-3 to the cipher text. The box is as shown in Table 5.2

Table 5.2

	A	B	C	D	E
A	T	H	U	R	S
B	D	A	Y	B	C
C	E	F	G	I	K
D	L	M	N	O	P
E	Q	V	W	X	Z

and HAPPY BIRTHDAY enciphers to

ABBBD EDEBC BDCDA DAAAB BABBB C

Since the transposition involves 5 numbers we write the cipher text into a box which has 5 columns. Note that since there are 26 letters in the cipher one column in the box needs to have 6 letters in it.The recipient of the message will need to know which, if any, are the 'long' columns so this will have been arranged beforehand. We will make the arbitrary assumption that the 'long' columns are formed in the order of the transposition. In the case of this example therefore the solitary 'long' column is the one headed '1', which is the second column from the left. Entering the cipher text into the transposition box we have Table 5.3.

Table 5.3

5	1	4	2	3
A	B	B	B	D
E	D	E	B	C
B	D	C	D	A
D	A	A	A	B
B	A	B	B	B
.	C	.	.	.

The text as finally transmitted is then

BDDAA CBBDA BDCAB BBECA BAEBD B

and the original digraph pairs have been broken up, making the cipher much harder to solve. The use of a 5×5 monograph–digraph substitution is still apparent but we can hide this quite simply by re-using the square to convert the digraphs back into monographs (Table 5.2), giving the new cipher monograph text

BLUAL BEACE DVM

and the method of encipherment is nicely hidden. It is, of course, essential to include a transposition before re-using the square since we will otherwise simply *decipher* the message!

MDTM ciphers

When used in this way the method is, strictly, what one might call a 'monograph–digraph–transposition–monograph' system, MDTM for brevity. The relationship between the individual letters of the plain and cipher texts is quite complex since each original letter is replaced by two letters each of which then separately combines with a letter from another pair to form a digraph, which is then re-converted into a monograph. Under certain circumstances this can produce a cipher text which looks bizarre when compared to the original plaintext, as the following shows.

Example 5.3

Use the MDTM method to encrypt the 25-letter alphabet (J omitted) using the normally ordered 5×5 square and the transposition 3-1-6-4-5-7-2-9-10-8.

Encryption

The 'plaintext' is

ABCDEFGHIKLMNOPQRSTUVWXYZ

which, using the 5×5 square, becomes

AAABA CADAE BABBB CBDBE CACBC CCDCE DADBD
CDDDE EAEBE CEDEE

Table 5.4

3	1	6	4	5	7	2	9	10	8
A	A	A	B	A	C	A	D	A	E
B	A	B	B	B	C	B	D	B	E
C	A	C	B	C	C	C	D	C	E
D	A	D	B	D	C	D	D	D	E
E	A	E	B	E	C	E	D	E	E

Writing this into the transposition rectangle, Table 5.4, we produce the 'digraph cipher' text

AAAAA ABCDE ABCDE BBBBB ABCDE ABCDE CCCCC
EEEEE DDDDD ABCDE

and then convert back to obtain the monograph cipher

AAAHU BOWGG BOVHU NNPZZ TTQHU

which one would not guess to be the encipherment of the 25-letter alphabet written in its normal order. This rather bizarre encryption is mainly due to the fact that we used the 5×5 alphabet square with no keyword followed by a transposition rectangle of width 10. In choosing the width of a transposition, multiples of the size of the alphabet square should be avoided.

Example 5.4 (A 1918 German High Command cipher)

Early in 1918 the German High Command began to use a cipher based upon a 5×5 square. The 25 letters of the alphabet were enciphered as digraphs using only the letters A, D, F, G and X. A modified form, using a 6×6 square, provided an additional 11 places and so enabled the users to include all 26 letters of the alphabet and the 10 numerals. The cipher was known as the 'ADFGVX cipher'. Each letter of the original message thus became a pair of letters which were then separated and transposed according to a key which changed daily. The cipher was not easy to solve but a

French cryptanalyst, Georges Painvin, worked out a method for doing it given a number of messages with identical plaintext beginnings or endings. Although only a few days were solved the number of messages on those days was high and their contents were particularly valuable. It is said that 'On one occasion the solution was so rapid that an important German operation disclosed by one message was completely frustrated' [5.1].

Example 5.5 (A Japanese naval cipher)
During World War II the Japanese Merchant Navy used a cipher, known as JN40, in which each Japanese *kana* symbol was replaced by two digits from a 10×10 square. The individual digits were then written into the *columns* of a rectangle, the ordering of the columns being given by a transposition. The digits were then taken out *row by row*, the ordering of the rows being given by a second transposition. The transpositions were changed every day. The cipher was solved in November 1942 when an operator left some details out of a message and then re-enciphered the full text using the same keys [5.2].

Problem 5.1
An MDTM system with keyword `ABSOLUTE` and transposition 3-1-5-2-4 has been applied to a message and the resulting cipher text is

```
CFIGS FLTBC XKEEA EBHTB GLDPI
```

Decrypt the message.

Digraph to digraph

Just as a simple substitution system replaces each individual letter of the alphabet by a single letter so one can construct a system in which every digraph is replaced by two letters. The 'obvious' way of doing this, as has been indicated before, is to have a list of all 676 ($=26 \times 26$) possible digraphs and their cipher equivalents e.g.

`AA = TK, AB = LD, AC = ER,..., ZX = DW, ZY = HB, ZZ = MS,`

but this involves having two lists, one for encryption and one for decryption, each 676 long, and although this would provide a better level of security than simple substitution it would be tedious to use. An alternative method is to use a digraph square which contains just the 25 letters of the alphabet, excluding `J` say, and form the cipher digraphs from the plaintext digraphs according to some rule. A nineteenth century system of this type is the method of

Playfair encipherment

The alphabet, ignoring J, is written into a 5×5 square in some order, either starting with a keyword, with the remaining letters following in order, or in a 'random' order. If a keyword has any repeated letters the second and later occurrences are ignored. Thus if TOMORROW is the keyword it would go into the square as TOMRW.

Encipherment is then carried out digraph by digraph according to the following rules.

(i) If the two letters of a digraph are at diagonally opposite corners of a rectangle in the 5×5 square they encipher to the pair at the other two corners. In the example below we adopt the convention that each letter is replaced by the corner letter in the same *row* as itself. The alternative convention is to replace each letter by the letter in the same *column* as itself and this is the usual practice with *double Playfair* ciphers, as we see below.

(ii) If the two letters are in the same column of the 5×5 square they encipher as the letters below them, where row 1 is considered to be below row 5 if necessary.

(iii) If the two letters are in the same row of the 5×5 square they encipher as the letters to the right of them, where column 1 is considered to be to the right of column 5 if necessary.

(iv) If the two letters are identical a 'dummy' letter, such as Q, is inserted between them.

(v) If necessary, a 'dummy' letter is inserted at the end of the plaintext.

Example 5.6

Encipher the message

SUPPORT NEEDED URGENTLY

using Playfair encipherment with the keyword WALKING.

Table 5.5

W	A	L	K	I
N	G	B	C	D
E	F	H	M	O
P	Q	R	S	T
U	V	X	Y	Z

The Playfair square is shown in Table 5.5. SU enciphers as PY since P is in the same row as S and Y is in the same row as U. The next pair PP will have to have a dummy inserted between them and so becomes two pairs, PQ and PO, which encipher as QR and TE respectively. Continuing in this way we find that the cipher text is

PYQRT ESPEP ONONX PNFDP KX

and is one letter longer than the original text because one dummy insertion was required.

Note that although the plaintext had another repeated-letter pair, EE in NEEDED, there was no need to insert a dummy letter because, at the point of encryption, the digraphs of the text split the EE apart, as NE followed by ED. Had we not already had to insert one dummy letter, between the PP in SUPPORT, the EE pair would not have been split and a dummy would then have had to be inserted. Likewise there was no need for a dummy letter at the end since the single dummy already inserted made the text length even.

Playfair decipherment

In some cipher systems decipherment involves precisely the same steps as encipherment but this is not so in others. In the Playfair system it is partly true but there is some asymmetry caused by the fact that where the two letters to be enciphered were in the same row or column the enciphered letters are obtained by moving to the *right* or *down* respectively. In deciphering therefore, when the two cipher letters fall in the same row or column the plaintext letters are obtained by moving to the *left* or *up* respectively. Where the two cipher letters are not in the same row or column the plaintext letters are those in the opposite corners of the rectangle as before.

Problem 5.2

A message has been enciphered using a Playfair cipher with the keyword RHAPSODY, the cipher text is

OXBGI HPEOK GHMTT ROIUE VGKGN C.

Decrypt the message.

Cryptanalytic aspects of Playfair

Playfair encipherment has certain properties that a cryptanalyst might exploit including the following.

(i) A letter cannot encipher to itself.

(ii) A letter can only encipher to one of 5 other letters which are the 4 other letters in its row and the letter below it in its column.

(iii) A letter is twice as likely to encipher to the letter immediately to its right than to any other letter. So, for example, in the square used in Example 5.6 (Table 5.5), if the second letter of a digraph beginning with M ***is not*** in the same row or column as M then M enciphers as E, F, H or O. If the second letter of the digraph ***is*** in the same ***row*** as M then M enciphers as O and if the second letter is in the same ***column*** as M then M enciphers as S. It follows that of the 24 possibilities for letters which can follow M in the plaintext (since J is not included and M would cause a dummy to be introduced) E, F, H, O, I, D, T and Z each cause O to occur as the cipher letter whereas the other 16 letters will cause E, F, H, and S each to occur four times. So in this case M will encipher to O twice as often as to any other letter and the same feature will hold for any other letter.

(iv) There is reciprocity between plain and cipher digraphs unless the letters are in the same row or column; that is, if CR, say, enciphers to PJ then PJ enciphers to CR and, furthermore, RC enciphers to JP and vice versa.

The usual method of attack on a Playfair cipher is via the digraphs. With a sufficiently long text a count of the cipher digraphs should reveal likely candidates for the cipher equivalents of high frequency plaintext digraphs such as TH, HE, IN and ER and since the reversals of two of these digraphs, HT and EH, have very *low* frequencies identification is made that much easier. Having identified the relative positions of some letters it may be possible to deduce which letters are in the keyword and then to reconstruct the Playfair square. An example of such a solution using a text of over eleven hundred digraphs is given in [5.3].

Double Playfair

Playfair encipherment was used by the British for some of their military ciphers up to and including the 1914–18 war and the Germans had considerable success in reading the messages. In World War II the German Army, knowing the weakness of the single Playfair cipher, used a double

Playfair system as their medium grade cipher until the autumn of 1944 after which it was replaced by a stencil cipher (see Chapter 7).

In a double Playfair system two 5×5 squares, placed side by side, are employed with the first letter of each plaintext digraph being located in the first, or left-hand, square and the second letter of the digraph located in the second, or right-hand, square. The corresponding cipher letters are obtained by the usual Playfair rules but, probably for ease of use, letters at the corners of a rectangle are replaced by the corresponding letters in the same *columns* as themselves, which means that a letter is always replaced by a letter from the same square as itself. Since the two letters use different squares they cannot be in the same column, but they can be in the same row. The alphabet in each square was in random order and was changed daily at midnight. To make the cipher more secure the message was written out in lines of fixed length and the pairs of letters to be enciphered were chosen *vertically* so that, if the length of the lines was 17, say, the letters being enciphered would be 17 apart in the original plaintext. The letter X was used as a word separator and, if necessary, to bring the text to an even length. Since the letters of a plaintext pair used two different squares there was no need to insert dummy letters between repeated letters. The system is illustrated by

Example 5.7
Use the Playfair squares in Table 5.6

Table 5.6

G	E	U	P	M	K	E	O	H	S
S	K	R	B	T	C	X	U	Z	F
C	Z	N	X	H	M	Q	B	R	W
O	Y	D	A	W	T	G	P	L	Y
L	F	V	I	Q	I	D	N	V	A

to encipher the message

OURXSITUATIONXISXDESPERATEXSENDXSUPPLIESXATXONCE

based upon a line length of 11 characters.

Encipherment
Since the message consists of 48 characters we will have four lines of 11 and two lines of 2 giving 24 pairs to be enciphered:

```
OURXSITUATI
ONXISXDESPE

RATEXSENDXS
UPPLIESXATX

ON
CE
```

The first vertical pair is OO; these letters are at the corners of a rectangle and so are replaced by the corresponding letters in their own columns: GP.

The second vertical pair is UN; they are replaced by VO.

The third vertical pair is RX; these are not at the corners of a rectangle since they are both in the second row of their boxes. Each is therefore replaced by the letter to its right giving BU as the cipher pair.

We continue in this way and produce the cipher text in the 11-character line format:

```
GVBIGBQPPWP
POUMFDXOYUD

BWWYIGURVAK
ZLUHMXKQYMU

SU
TQ
```

(Verification is left as an exercise for the reader.)

These would now be transmitted, line by line in order, as 5-letter groups. The last group might, or might not, be padded out with random characters:

```
GVBIG BQPPW PPOUM FDXOY UDBWW YIGUR VAKZL
UHMXK QYMUS UTQ
```

As the example shows encipherment is a rather laborious business and the German cipher operators frequently made mistakes, sometimes using the wrong squares, and this led to requests for repeats which helped the cryptanalysts considerably. There were also quite a lot of standard phrases in the messages and the messages were deciphered regularly. For further details see [5.4].

6

Codes

Characteristics of codes

As was mentioned in Chapter 1, the distinction between codes and ciphers is not always clear, but one might reasonably say that whereas most codes tend to be *static* most ciphers are *dynamic*. That is to say that a letter or phrase enciphered simply by means of a code will produce the same cipher each time the code is used, whereas a letter or phrase enciphered by a cipher system will generally produce different cipher text at different times. This is because most cipher systems involve one or more parameters, such as keywords or, as we shall see later, wheel settings, which are changed at regular or irregular intervals and so cause the cipher outputs from the same plaintext to be different. The basic mechanism, or algorithm, for generating the cipher doesn't change, but the parameters do. In general, a code has no such parameters though the entire code may itself be changed, in which case it becomes a different code. In practice this is achieved by issuing a new code-book every now and then. Using this criterion the Julius Caesar cipher would be classed as a code, because the encipherment of a fixed letter across many messages is invariably the same. We can, however, say that there is a parameter associated with Julius Caesar ciphers, namely the shift, which gives us 25 different ciphers and if the value of the shift is somehow incorporated in the message, i.e. in the indicator, the Julius Caesar system can reasonably be considered as a cipher, not as a code.

Example 6.1

Although there are much earlier examples of codes the one devised by Samuel Morse (1791–1872) in 1832 for the purpose of transmitting messages by telegraphy is probably the best-known. In this code the letters of the alphabet are represented by up to *four* 'dots' and 'dashes', the digits 0 to 9 by *five* and certain punctuation symbols by *six*. To transmit a dot the telegraph key is depressed for about one 24th of a second; for a dash the key is depressed for about one 8th of a second; the interval between the components of a letter is the same as that for a dot and the interval between letters is equal to that for a dash. The Morse code was designed so that the most frequent letters in English had shorter transmission times than the less frequent letters. Thus E was represented by a single dot and T by a single dash whereas J required four symbols, dot dash dash dash. The reason for this was to try to minimise the time required to transmit a message. The international wireless version of the letters of the Morse code is shown in Table 6.1.

Table 6.1 *Morse code*

A ·—	E ·	I ··	M ——	Q ——·—	U ··—	Y —·——
B —···	F ··—·	J ·———	N —·	R ·—·	V ···—	Z ——··
C —·—·	G ——·	K —·—	O ———	S ···	W ·——	
D —··	H ····	L ·—··	P ·——·	T —	X —··—	

The Morse code was not, of course, designed to protect the *secrecy* of a message but merely to provide a means for transmitting it efficiently. A good wireless operator using this code would be able to transmit about 30 average words per minute. As was mentioned in Chapter 1, there are other codes which are designed to ensure the accuracy of messages or data rather than to preserve the secrecy of their contents. Among such codes are those used to transmit data from spacecraft or to store data in computer-readable form. If secrecy is not needed the details of the code will usually be available to anyone who wants them. If secrecy as well as accuracy is required the details may not be made public and some form of encryption of the data will also be applied.

One-part and two-part codes

Most codes involve the use of a ***code-book***, which may contain thousands of ***code groups***. A code used by the military would typically represent letters, numbers or phrases by code groups consisting of four or five letters or digits. It is not necessary that all the code groups contain the same number of symbols; the famous Zimmermann telegram of January 1917, which was deciphered by British cryptanalysts and which was a major factor in America's decision to enter the War, had a mixture of code groups of both four and five digits [6.1]. The main advantage of a code is that it can provide many code groups; up to 10 thousand for a four-digit code and nearly 12 million for a five-letter code. The disadvantages are that (i) it is necessary for the users to carry the code book(s) with them and (ii) if the enemy acquire a copy of the book, either by capture or by breaking the code, reading of future messages is straightforward. For these reasons codes are more likely to be employed by embassies or large military units, such as ships, than by individuals.

Breaking of a code is made very much easier if the code is a ***one-part code*** which means that the same code-book is used for both encipherment and decipherment. If this is so the code groups for words or phrases which are close in a dictionary will be close to each other numerically. Thus a section of a four-digit one-part codebook might look like Table 6.2.

Table 6.2

A	0001
ABLE	0013
AFTER	0023
AM	0051
AN	0075
AND	0078
ANY	0081
AS	0083
ASK	0091
AT	0097

Not all of the 10 thousand possible code groups would normally be used. Gaps would probably be left which would allow other words or phrases to be inserted later if desired.

Table 6.3

A	5832
ABLE	2418
AFTER	6941
AM	9075
AN	6948
AND	4729
ANY	8532
AS	4271
ASK	2163
AT	1894

From a cryptographic point of view a one-part code offers the cryptanalyst too great an advantage by enabling him to guess at the meanings of as yet unidentified code groups simply by looking for plausible words in a dictionary which are close to identified words. This weakness can be removed by making the numerical ordering of the code groups unrelated to the alphabetical or numerical order of the codewords. We then have a ***two-part code*** and the users need two code-books, one for encipherment and one for decipherment. The codewords above might then appear like Table 6.3 in the encipherment book, whilst the decipherment book might begin as in Table 6.4 and so on.

Table 6.4

0005	TOMORROW
0009	ATTACK
0014	COME

In all cases, it is likely that very common codewords would be allocated more than one code group and the users instructed to use each of the alternatives in a 'random' manner.

Although codes which have not been subjected to further encipherment do not offer a high level of security they *have* been used in wartime; the Italian Navy used a one-part code, known as *Mengarini* [6.2], for some very low grade messages, and the Japanese Navy used a two-part code, known to them as *OTSU* and to the British as JN4 [6.3], during the Second World War. A somewhat different code using two letters followed by four digits was used by U-boats of the German Navy to report their positions in the Atlantic and to receive instructions for attacking Allied convoys. The letters were subjected to digraph substitution tables and the digits could also be modified [6.4].

Code plus additive

No matter how many code groups a code contains, a cryptanalyst, given enough messages, will eventually find certain groups occurring more than once, even when the same plaintext word or phrase has several alternative code groups allocated to it. Also, if the code-book is captured by the enemy, decryption of all messages becomes trivial. To overcome these weaknesses the code groups themselves are usually enciphered. A standard way of doing this is to apply an ***additive key*** to the code groups using *non-carrying*, or *modular, addition*. Although this has been mentioned before, to remind ourselves how it is done, let us look at an example. Suppose that we have the code group 6394 and that the key to be applied to it is 2798; then the code group is written down, the key placed directly underneath it and corresponding digits are added *without carrying* so that when we add the last digits of the code and key, 4 and 8, the sum is written as 2, not 12 (that is: we are adding digit by digit (mod 10)). So we have

Code group	6394
Key	2798
Sum	8082

and the cipher text is 8082. The key would not be the same for the other groups since in practice the key either would not repeat at all or, if it did repeat, would only do so after many digits. Since *encryption* involves *adding* the key to the code groups the person receiving the message would have to *subtract* the key digit by digit (mod 10) from the cipher in order to recover the code groups and so *decipher* the message; thus:

Cipher	8082
Key	2798
Code group	6394

Obviously the code groups are now disguised, and the security of the system is substantially increased provided that the key does not repeat for a sufficiently long period. The question of how to produce sequences of digits which do not repeat until many thousands have been generated is one of considerable interest to mathematicians and cryptographers, and we consider it more fully in Chapter 8. In the mean time, by way of illustration, here is a very simple method, which generates a sequence which repeats after 60 digits.

Example 6.2

Generate a sequence of digits (mod 10) by starting with the digits 3 and 7 and forming each new digit by adding together the two previous digits (mod 10).

Solution

The sequence starts

3 7

so the next digit is $(3+7)=10$ which is 0 (mod 10) and the 4th digit is $(7+0)=7$ and so the 5th digit is $(0+7)=7$. Continuing in this way we find that the sequence which is generated is

3 7 0 7 7 4 1 5 6 1 7 8 5 3 8 1 9 0 9 9 8 7 5 2 7 9 6 5 1 6 7 3 0 3 3 6 9 5 4
9 3 2 5 7 2 9 1 0 1 1 2 3 5 8 3 1 4 5 9 4 3 7 0...

and the sequence begins to repeat after 60 digits, as indicated by the underlining of the last 3 digits, which are the same as the first 3. Since each digit is the sum (mod 10) of the previous 2 the key will begin to repeat when 2 digits occur which have already occurred *in the same order* earlier in the sequence. It follows that any sequence generated (mod 10) in this way cannot have a period longer than 100, since there are only 100 pairs of digits (mod 10). The sequence of the example, with a period of 60, is the longest available sequence in this case. Had we begun with the first 2 terms both equal to 0 we would have produced an all-zero sequence which, if used as a key, would leave the code groups unaltered.

Although it is the longest available by this simple method the key sequence of the example unfortunately has certain numerical properties which make it undesirable from a cryptographic point of view. One that is particularly bad is that two-thirds of the digits are *odd* and only one-third are *even*, instead of there being approximately equal numbers of each. This is because the sequence has a very simple *odd–even* pattern, as we can see immediately, viz:

odd odd even odd odd even....

Another property is that double digits (77, 99, 33 and 11) occur regularly, 15 places apart. This particular sequence is very well-known and is a special case of what is probably the most intensively studied sequence in mathematics. The usual form of it starts with 0 and 1 as the first 2 values and continues as in the example but without reduction (mod 10), i.e. the terms are added normally. The sequence then begins

0 1 1 2 3 5 8 13 21 34 55 89 144 233 377 610 987 1597...

The terms grow at a tremendous rate (the mathematical expression is that they grow 'exponentially'); each term after the 5th is more than 1.5 times bigger than its predecessor and so, for example, the 100th term in the sequence contains 21 digits. If we reduce all these numbers (mod 10), which is the same as replacing each of them by its last digit, we get

0 1 1 2 3 5 8 3 1 4 5 9 4 3 7 0 7 7...

which is the same as the sequence in the example but starting at position 48.

The sequence beginning 0 1 1... is known as the Fibonacci sequence; it was introduced into mathematics by Leonardo of Pisa, also known as Fibonacci, in the thirteenth century. For further information on this famous sequence see M5.

Problem 6.1

(1) Generate the key (mod 10) as above but starting with 0 and 2 as the first 2 terms. What is the period and why would this key be rejected by a cryptographer?

(2) What is the period when we start with 1 and 3 as the first 2 terms?

The Fibonacci sequence is the simplest of a class of sequences that can be used for the generation of keys for use in cryptography, although whether any particular sequence provides 'good' keys, in that all possible key values are equally likely to occur, is a question which can only be answered by using some advanced mathematics [1.2, 1.3]. A fairly obvious generalisation of the Fibonacci sequence is obtained by forming each new term by adding together (mod 10) the 3 preceding terms which may produce sequences of longer cycle length thus:

Example 6.3

Starting with 0, 1, 1 as the first 3 terms generate the sequence obtained (mod 10) by adding together the previous 3 terms at each stage.

Solution

The sequence begins 0 1 1 2 4 7 3 4 4 1... and repeats after 124 terms. The frequencies of the individual digits are slightly non-uniform; each should occur 12 or 13 times but 3 occurs only 6 times whereas 4 and 9 both occur 18 times. Verification of these facts is left to the reader.

If we choose three different terms for starting the sequence we may get shorter cycle lengths: 0, 1, 2 produces a cycle of length 62 whilst 0, 5, 0 would be a very poor choice since it generates a sequence of cycle length just 2.

This method can be used to generate keys to any modulus. For example, the sequence which starts 0, 1, 1 to moduli 2, 3, 5 and 7 produces cycles of length 4, 13, 31 and 48 respectively. From a purely mathematical point of view this is interesting but, in general, cryptologists would probably use 2, 10 or 100 as the modulus.

If we generate the same sequence (mod 100) it begins

00 01 01 02 04 07 13 24 44 81 49 74 ...

and is found to repeat after 1240 terms.

It is possible to modify the Fibonacci sequence so that the odd–even ratio is somewhat reduced, making it slightly better for encryption. A modest step in this direction is shown by

Example 6.4

Generate 20 terms of the Fibonacci sequence (mod 100) starting with 13 and 21 as the first 2 terms then interchange the second and third digits in each group of four to give 20 terms of a two-digit key stream.

Solution

The first 20 terms of the Fibonacci Sequence (mod 100) starting with 13 and 21 are

13 21 34 55 89 44 33 77 10 87 97 84 81 65 46 11 57 68 25 93

We interchange the second and third digits in each group of four –

12 31 35 45 84 94 37 37 18 07 98 74 86 15 41 61 56 78 29 53

– and this is the resultant key. The bias of *odd* : *even* numbers has now been reduced (from 2 : 1 to about 7 : 5) and the key, though still unsatisfactory, is stronger for that.

Problem 6.2

A two-digit code represents the letters of the alphabet as follows:

A = 17, B = 20, C = 23, ..., Z = 92,

each number being 3 more than the one before it. A message is then enciphered using this code and the additive key (12 31 35...) obtained in the example above, the addition being digit by digit with no carrying. The resultant cipher text is

86 69 42 19 60 35 08 13 76 48 23 02 50 91.

Decrypt the message.

7

Ciphers for spies

A spy operating in country X on behalf of country Y has the problem of communicating with his controller in such a way as to protect both himself and the contents of his messages. No matter how he sends his messages they will have to be 'modified' somehow so that their true meaning is hidden from anyone but the intended recipient. There are methods, such as the use of microdots or 'invisible' ink, which do not, *per se*, involve encipherment although some 'modification' of the text even in such cases would probably be used to provide extra security. When we say that a text has been 'modified' we do not necessarily mean that it has been enciphered but that the 'secret' text is not simply sent in an unaltered form: it might, for example, be hidden inside an apparently innocuous message.

Hiding a secret text inside an innocuous one has the advantage that, being apparently unenciphered, it will not automatically attract the interest of unintended recipients or interceptors, such as the security forces of country X. A disadvantage is that it may not be too easy to construct a realistic non-secret text in which to embed it. Here is a simple illustration.

Example 7.1 ('Part of a letter from Agent 63')
As I was walking through the centre of town yesterday morning at about eleven thirty I chanced to see Ron Kingston. He was alone, driving a newish-looking ultramarine car, a Ford Escort. Previously he's had only second-hand cars, not often less than three years old. Perhaps he has had an inheritance from some rich relative who has recently died?

Secret message
If we take the first letter of the message and the first letter of every fourth word after that, taking hyphenated words as two separate words, we get

```
ATTACKDUEONTHIRD
```

Stencil ciphers

The example above is a very simple, and insecure, case of a stencil cipher. In such a cipher certain letters on a page are part of the secret message and all the other letters are merely 'fillers' which are used to compose a mundane-looking communication. The 'stencil' in the example is far too regular to be satisfactory; a more suitable stencil would use letters which are separated from each other by irregular intervals and which are not necessarily the first letters of words. To add further security the letters of the secret text would probably not occur in their correct order in the overall text. If the overall text is typed in a regular format the sender and recipient may have identical cards with holes punched in them at the positions of the letters of the secret message. Hence the name of this type of cipher. Each hole would have a number underneath it giving the position of its corresponding letter in the message.

Such a cipher would be much harder to solve unless the same stencil was used repeatedly in which case, given enough messages, it might be possible for a cryptanalyst to recover part of the stencil and from that gradually to recover the rest. If, however, the stencil changed regularly it would be extremely difficult, if not impossible without further information, to solve the system. To get some idea of the difficulty consider this.

Example 7.2 (stencil cipher)
Text of message:

```
Some of Shakespeare's plays such as
     10    6                  8   1
Anthony and Cleopatra are performed
  2      11   9   3         15     7
less frequently than others, Macbeth
            12    4            5   13
King Lear and Hamlet, in particular.
 14         16
```

The numbers show where the holes in the stencil were placed and give the order in which the visible letters are to be read; where a two-digit number is needed the letter above the first digit is the one where the hole occurs; the letter has been underlined for clarity, though of course it would not have been underlined in the message as sent! Thus the message reads, after insertion of spaces,

```
ATTACK DUE ON THIRD
```

as before.

If a stencil is used several times it might be possible for a cryptanalyst to discover where the holes are placed, and in what order they are used, but it would depend on whether any likely plaintext phrases or words were available and on the number and length of the messages. If the stencil was changed after every page it would be impossible to read the messages unless there was some relationship between the make-up of a message stencil and its successor, for without knowing the stencil many possible messages might be found within a page as is illustrated by

Problem 7.1

Verify that all but one of the following messages can be found within the text of the example above and so are 'possible solutions' given an appropriate stencil:

(1) MERRY CHRISTMAS;
(2) COME AT ONCE;
(3) GO AWAY QUICKLY;
(4) THE AUTHOR OF OTHELLO IS BACON.

Hundreds of other 'possible solutions' could be found within the text since it contains over a hundred letters and any anagram of any subset could be picked out with an appropriate stencil. Without further information such as, for example, that a stencil will not have more than one hole in any row or column, such a message is 'unbreakable' since many solutions are possible. The same situation can apply to other cipher systems where insufficient material is available to provide a unique solution. Even a simple substitution system is unbreakable if it is used only once for a single short message. In the extreme case of a 'one-time pad' the system is unbreakable no matter how many messages of any lengths are sent, as we shall see later. If the same system (simple substitution, transposition, stencil) is used more than once it may cease to be 'unbreakable'; even a 'one-time pad' may lose its security *if the same pad is used twice*.

Book ciphers

A spy must avoid arousing suspicion and so any cipher equipment he has in his house must not be obvious. Even a single stencil might appear suspicious to an investigator and a stack of stencils could be incriminating. Likewise a spy would be unlikely to use a code if it meant having to have a large code-book in the house. A cipher that requires no unusual equipment is therefore a very attractive proposition so far as a spy is concerned and a book cipher is precisely that; all that is required is a book on any topic which does not employ non-Latin alphabetic characters. The book could, for example, be an English novel or a biography or historical work but probably not one dealing with organic chemistry.

Using a book cipher

In order to use a book cipher it is necessary to be able to 'add and subtract' pairs of letters of the alphabet. This is done, as explained in Chapter 1, by numbering the letters of the alphabet A=0, B=1, C=2,..., Z=25 and adding or subtracting (mod 26) and re-converting the answers to letters. Since this is a tedious process it is better to make up tables once and for all and look up the result of adding or subtracting in the appropriate table, but to show how to do it without such tables let us carry out the process on a few letters.

Example 7.3

Convert the alphabet to numbers beginning with A=0, B=1 etc. Then 'add' together the two texts below (mod 26) and re-convert the resulting numbers to letters.

Text 1 THEXCURFEWXTOLLSX

Text 2 ONCEXUPONXAXTIMEX

Solution

We repeat Table 1.1 as Table 7.1.

Table 7.1

A	B	C	D	E	F	G	H	I	J	K	L	M	N	O	P	Q	R	S	T	U	V	W	X	Y	Z
00	01	02	03	04	05	06	07	08	09	10	11	12	13	14	15	16	17	18	19	20	21	22	23	24	25

We first convert the texts to numbers by using the table:

Text 1	T	H	E	X	C	U	R	F	E	W	X	T	O	L	L	S	X
	19	7	4	23	2	20	17	5	4	22	23	19	14	11	11	18	23
Text 2	O	N	C	E	X	U	P	O	N	X	A	X	T	I	M	E	X
	14	13	2	4	23	20	15	14	13	23	0	23	19	8	12	4	23

Now we add them and then reduce them (mod 26):

	19	7	4	23	2	20	17	5	4	22	23	19	14	11	11	18	23
	14	13	2	4	23	20	15	14	13	23	0	23	19	8	12	4	23
Sum:	33	20	6	27	25	40	32	19	17	45	23	42	33	19	23	22	46
(mod 26):	7	20	6	1	25	14	6	19	17	19	23	16	7	19	23	22	20

Finally we convert the numbers back into letters, using the table:

H U G B Z O G T R T X Q H T X W U

and this is the cipher text that would be sent. The recipient would, of course, need to *subtract* Text 2 from the cipher (mod 26) in order to recover Text 1, viz:

Cipher	H	U	G	B	Z	O	G	T	R	T	X	Q	H	T	X	W	U
Text 2	O	N	C	E	X	U	P	O	N	X	A	X	T	I	M	E	X
Convert	7	20	6	1	25	14	6	19	17	19	23	16	7	19	23	22	20
	14	13	2	4	23	20	15	14	13	23	0	23	19	8	12	4	23

Subtract (mod 26), i.e if the result is negative, add 26:

19 7 4 23 2 20 17 5 4 22 23 19 14 11 11 18 23

Re-convert to letters:

T H E X C U R F E W X T O L L S X

which is Text 1, the original 'message'.

Obviously it would be a very tedious and error-prone process to have to convert the texts to numbers, add them, subtract 26 where necessary, and re-convert to letters every time a message was to be enciphered so it is very worthwhile having two tables, one for enciphering and one for deciphering, from which the result of applying these processes can be read off immediately. Experienced users would not need such tables since they would soon learn to 'add the letters' at sight but for others the tables save a lot of time and effort. They are given in Tables 7.2 and 7.3. Notice that in the *encipher* table (Table 7.2) it makes no difference whether we call the

Table 7.2 *Encipher table for a book cipher*

	Text 1																									
Text 2	A	B	C	D	E	F	G	H	I	J	K	L	M	N	O	P	Q	R	S	T	U	V	W	X	Y	Z
A	A	B	C	D	E	F	G	H	I	J	K	L	M	N	O	P	Q	R	S	T	U	V	W	X	Y	Z
B	B	C	D	E	F	G	H	I	J	K	L	M	N	O	P	Q	R	S	T	U	V	W	X	Y	Z	A
C	C	D	E	F	G	H	I	J	K	L	M	N	O	P	Q	R	S	T	U	V	W	X	Y	Z	A	B
D	D	E	F	G	H	I	J	K	L	M	N	O	P	Q	R	S	T	U	V	W	X	Y	Z	A	B	C
E	E	F	G	H	I	J	K	L	M	N	O	P	Q	R	S	T	U	V	W	X	Y	Z	A	B	C	D
F	F	G	H	I	J	K	L	M	N	O	P	Q	R	S	T	U	V	W	X	Y	Z	A	B	C	D	E
G	G	H	I	J	K	L	M	N	O	P	Q	R	S	T	U	V	W	X	Y	Z	A	B	C	D	E	F
H	H	I	J	K	L	M	N	O	P	Q	R	S	T	U	V	W	X	Y	Z	A	B	C	D	E	F	G
I	I	J	K	L	M	N	O	P	Q	R	S	T	U	V	W	X	Y	Z	A	B	C	D	E	F	G	H
J	J	K	L	M	N	O	P	Q	R	S	T	U	V	W	X	Y	Z	A	B	C	D	E	F	G	H	I
K	K	L	M	N	O	P	Q	R	S	T	U	V	W	X	Y	Z	A	B	C	D	E	F	G	H	I	J
L	L	M	N	O	P	Q	R	S	T	U	V	W	X	Y	Z	A	B	C	D	E	F	G	H	I	J	K
M	M	N	O	P	Q	R	S	T	U	V	W	X	Y	Z	A	B	C	D	E	F	G	H	I	J	K	L
N	N	O	P	Q	R	S	T	U	V	W	X	Y	Z	A	B	C	D	E	F	G	H	I	J	K	L	M
O	O	P	Q	R	S	T	U	V	W	X	Y	Z	A	B	C	D	E	F	G	H	I	J	K	L	M	N
P	P	Q	R	S	T	U	V	W	X	Y	Z	A	B	C	D	E	F	G	H	I	J	K	L	M	N	O
Q	Q	R	S	T	U	V	W	X	Y	Z	A	B	C	D	E	F	G	H	I	J	K	L	M	N	O	P
R	R	S	T	U	V	W	X	Y	Z	A	B	C	D	E	F	G	H	I	J	K	L	M	N	O	P	Q
S	S	T	U	V	W	X	Y	Z	A	B	C	D	E	F	G	H	I	J	K	L	M	N	O	P	Q	R
T	T	U	V	W	X	Y	Z	A	B	C	D	E	F	G	H	I	J	K	L	M	N	O	P	Q	R	S
U	U	V	W	X	Y	Z	A	B	C	D	E	F	G	H	I	J	K	L	M	N	O	P	Q	R	S	T
V	V	W	X	Y	Z	A	B	C	D	E	F	G	H	I	J	K	L	M	N	O	P	Q	R	S	T	U
W	W	X	Y	Z	A	B	C	D	E	F	G	H	I	J	K	L	M	N	O	P	Q	R	S	T	U	V
X	X	Y	Z	A	B	C	D	E	F	G	H	I	J	K	L	M	N	O	P	Q	R	S	T	U	V	W
Y	Y	Z	A	B	C	D	E	F	G	H	I	J	K	L	M	N	O	P	Q	R	S	T	U	V	W	X
Z	Z	A	B	C	D	E	F	G	H	I	J	K	L	M	N	O	P	Q	R	S	T	U	V	W	X	Y
	A	B	C	D	E	F	G	H	I	J	K	L	M	N	O	P	Q	R	S	T	U	V	W	X	Y	Z

message text 'Text 1' and the key text 'Text 2' or vice versa since adding the two texts gives the same result either way ('addition is commutative' is the mathematical phrase to describe this), but in the *decipher* table (Table 7.3) the cipher and the key must be correctly identified since to get the plaintext we must *subtract* the key from the cipher and *not* vice versa. This is evident when we recall that

to encipher: cipher = key + text

and so

to decipher: text = cipher – key

where the additions and subtractions are carried out (mod 26) of course.

Table 7.3 *Decipher table for a book cipher*

	Key text																									
Cipher	A	B	C	D	E	F	G	H	I	J	K	L	M	N	O	P	Q	R	S	T	U	V	W	X	Y	Z
A	A	Z	Y	X	W	V	U	T	S	R	Q	P	O	N	M	L	K	J	I	H	G	F	E	D	C	B
B	B	A	Z	Y	X	W	V	U	T	S	R	Q	P	O	N	M	L	K	J	I	H	G	F	E	D	C
C	C	B	A	Z	Y	X	W	V	U	T	S	R	Q	P	O	N	M	L	K	J	I	H	G	F	E	D
D	D	C	B	A	Z	Y	X	W	V	U	T	S	R	Q	P	O	N	M	L	K	J	I	H	G	F	E
E	E	D	C	B	A	Z	Y	X	W	V	U	T	S	R	Q	P	O	N	M	L	K	J	I	H	G	F
F	F	E	D	C	B	A	Z	Y	X	W	V	U	T	S	R	Q	P	O	N	M	L	K	J	I	H	G
G	G	F	E	D	C	B	A	Z	Y	X	W	V	U	T	S	R	Q	P	O	N	M	L	K	J	I	H
H	H	G	F	E	D	C	B	A	Z	Y	X	W	V	U	T	S	R	Q	P	O	N	M	L	K	J	I
I	I	H	G	F	E	D	C	B	A	Z	Y	X	W	V	U	T	S	R	Q	P	O	N	M	L	K	J
J	J	I	H	G	F	E	D	C	B	A	Z	Y	X	W	V	U	T	S	R	Q	P	O	N	M	L	K
K	K	J	I	H	G	F	E	D	C	B	A	Z	Y	X	W	V	U	T	S	R	Q	P	O	N	M	L
L	L	K	J	I	H	G	F	E	D	C	B	A	Z	Y	X	W	V	U	T	S	R	Q	P	O	N	M
M	M	L	K	J	I	H	G	F	E	D	C	B	A	Z	Y	X	W	V	U	T	S	R	Q	P	O	N
N	N	M	L	K	J	I	H	G	F	E	D	C	B	A	Z	Y	X	W	V	U	T	S	R	Q	P	O
O	O	N	M	L	K	J	I	H	G	F	E	D	C	B	A	Z	Y	X	W	V	U	T	S	R	Q	P
P	P	O	N	M	L	K	J	I	H	G	F	E	D	C	B	A	Z	Y	X	W	V	U	T	S	R	Q
Q	Q	P	O	N	M	L	K	J	I	H	G	F	E	D	C	B	A	Z	Y	X	W	V	U	T	S	R
R	R	Q	P	O	N	M	L	K	J	I	H	G	F	E	D	C	B	A	Z	Y	X	W	V	U	T	S
S	S	R	Q	P	O	N	M	L	K	J	I	H	G	F	E	D	C	B	A	Z	Y	X	W	V	U	T
T	T	S	R	Q	P	O	N	M	L	K	J	I	H	G	F	E	D	C	B	A	Z	Y	X	W	V	U
U	U	T	S	R	Q	P	O	N	M	L	K	J	I	H	G	F	E	D	C	B	A	Z	Y	X	W	V
V	V	U	T	S	R	Q	P	O	N	M	L	K	J	I	H	G	F	E	D	C	B	A	Z	Y	X	W
W	W	V	U	T	S	R	Q	P	O	N	M	L	K	J	I	H	G	F	E	D	C	B	A	Z	Y	X
X	X	W	V	U	T	S	R	Q	P	O	N	M	L	K	J	I	H	G	F	E	D	C	B	A	Z	Y
Y	Y	X	W	V	U	T	S	R	Q	P	O	N	M	L	K	J	I	H	G	F	E	D	C	B	A	Z
Z	Z	Y	X	W	V	U	T	S	R	Q	P	O	N	M	L	K	J	I	H	G	F	E	D	C	B	A
	A	B	C	D	E	F	G	H	I	J	K	L	M	N	O	P	Q	R	S	T	U	V	W	X	Y	Z

Problem 7.2

A message has been enciphered using a book cipher. The book used was *The Poems of Rupert Brooke* and the key for the message was the passage beginning

STANDSXTHEXCHURCHXCLOCKXATXTENXTOXTHREE

The cipher text was

LAEKV MPILG QZOUJ ZTLXP RZDLX EFOIE MHCIQ

Decrypt the message.

Letter frequencies in book ciphers

The frequencies of the letters of the alphabet in the cipher produced by a book cipher when the key is a passage of English text will be different from those of unenciphered English. Whilst it will certainly be true that some letters, such as those letters used for 'space', E and T, will no longer occur very much more frequently than others such as Z or J, it is also true that the letters will not be equally represented. It is possible to estimate the frequencies of the cipher letters and we find that although the frequencies of the individual letters do not vary so much as they do in samples of normal (unenciphered) English they are still by no means uniform and observation of this type of variation would alert the cryptanalyst to the possibility that a book cipher was being used. This variation can be seen in Table 7.4. In the left-hand column are the frequencies of the 26 letters of the alphabet, and a 27th 'letter' which covers all punctuation marks and 'space', as they occur in a typical sample of normal English and, in the right-hand column, the frequencies of the same letters as they are predicted to occur in a passage enciphered by a book cipher using English texts. The sample size in both cases is 1000 so that 'on average' each letter should occur about 37 times. It will be seen that this is very far from being the case in the unenciphered text, and even in the enciphered text there is considerable variation in the frequencies, though not a lot more than we would expect at random (for further comments see M6).

A book cipher might be regarded as an extreme case of a Vigenère cipher in which the key length is the same as the length of the message itself. Evidently a book cipher ought to be more secure than a Vigenère since the latter uses a key of fixed length.

An alternative to using a key which is an English text is to use a key in a different language but which uses an alphabet of not more than 26 letters. Diacriticals, such as accents or umlauts, would be ignored if French or German were used. This would make life somewhat harder for the cryptanalyst at least until he realised what was going on.

Solving a book cipher

Assuming that a cryptanalyst has realised that the cipher text is the result of enciphering an English text with an English book used as the key how might he go about trying to solve it? Although the letter frequencies in

Table 7.4 *Letter frequencies in unenciphered English and in book cipher*

	Frequency (per 1000)	
Letter	Normal English	Book cipher
A	64	31
B	14	30
C	27	29
D	35	55
E	100	40
F	20	32
G	14	39
H	42	46
I	63	34
J	3	26
K	6	36
L	35	37
M	20	45
N	56	38
O	56	28
P	17	26
Q	4	37
R	49	46
S	56	52
T	71	38
U	31	26
V	10	39
W	18	34
X	3	32
Y	18	25
Z	2	52
Space etc.	166	47

Source: The data in the left-hand column are quoted in [1.2, Appendix 2]; the data in the right-hand column were derived mathematically from those in the left-hand column (for the details see M6).

the cipher might have helped him to the conclusion that a book cipher is being used, the individual letter frequencies will not be so different from random that they will be of much help, nor will the digraphs etc. There is, however, an attack to which a book cipher is vulnerable: it might be called 'crib-dragging'. Suppose that either the message or the key text contains some common English word such as THE. This is added to 3 letters of another English word in the other text to produce 3 letters of the cipher.

If then we try *subtracting* THE from the cipher at every possible position and look at the trigraphs obtained we might find plausible looking parts of English words which we might then be able to complete, thus adding a few more letters in the other text, preceding or following THE. Other common trigraphs can be tried and so the two texts might begin to 'unravel', so to speak. If X is used as a separator we can extend THE to THEX, or possibly even to XTHEX, although in doing so we might fail to pick up a word such as THERE. Even a short word, such as A, might be helpful if it occurs as XAX.

If some unusual words can be found in the key text it might be possible to deduce the type of book being used and even to identify the book, which would make subsequent cryptanalysis much easier. In practice we might at first only be able to recover occasional words or parts of words in the two texts but even partial recovery could be informative and subsequent messages might provide further useful 'cribs'. As an indication of the method, on a small scale (just 50 letters):

Example 7.4

The following 10 groups have been enciphered using a book cipher. Use the technique of crib-dragging to try to recover the texts of the key and message.

```
FLIQT NYQFK VACEH UCUAC MOXRG EYYQJ BNOEQ
FJXUL ILREJ ATVQB
```

We try some common words as possible cribs, THE being the most obvious. Crib-dragging is undeniably tedious since we must try the crib at all possible positions of the cipher text. Since any of the letters that we try, T, H and E in this case, may occur in other cribs we might save some effort overall if we first see what the resultant plain letter would be if we assume that T is the letter to be subtracted from the cipher at each position, then if H is the letter and finally if E is the letter. If we now write out the three resulting streams of 'putative plaintext' with the first line (corresponding to T) offset two places to the right and the second line (the 'H' line) offset one place to the right relative to the third ('E') line then any possible 'words' will appear as three letters in a vertical line; thus:

```
Cipher     FLIQT NYQFK VACEH UCUAC MOXRG EYYQJ BNOEQ FJXUL ILREJ ATVQB
Tline      MSPXA UFXMR CHJLO BJBHJ TVEYN LFFXQ IUVLX MQEBS PSYLQ HACXI
Hline     YEBJMG RJYDO TVXAN VNTVF HQKZX RRJCU GHXJY CQNEB EKXCT MOJU
Eline    BHEMPJU MBGRW YADQY QWYIK TNCAU UMFXJ KAMBF TQHEH NAFWP RMX
```

There are a few plausible looking trigraphs such as

MEE at position 1,
ROW at position 10,
ONY at position 15,
BEE at position 39,
PEN at position 41.

We now try extending the crib from THE to THEX to see if it produces a plausible tetragraph in any of these five cases. We insert X at positions 4, 13, 18, 42 and 44; the cipher letters at these positions are Q, C, U, L and L respectively and we decrypt these by 'subtracting' X from them, which means in effect moving each of these letters three places forward in the alphabet, thus giving the decrypts T, F, X, O and H so that the tetragraphs produced are:

MEET at position 1,
ROWF at position 10,
ONYX at position 15,
BEEO at position 39,
PENH at position 41.

The first of these looks particularly promising so we immediately investigate that further, keeping the others for analysis later. Since the first 'hit' of THE was at position 1 we examine the cipher text immediately following, the first 10 places initially. What we tentatively have is

Cipher	FLIQT NYQFK
Text1	THEX.
Text2	MEET.

The first word in Text 2 might be MEET in which case it should be followed by X, as the separator, or it might be a longer word such as MEETINGX. In the first case the fifth letter of Text 1 would be W; in the second case letters 5, 6, 7 and 8 of TEXT 1 would be (T−I), (N−N), (Y−G) and (Q−X), i.e L, A, S and T, which looks extremely good since it implies that Text 1 begins

THE LAST.

We would now expect that LAST would be followed by X in which case the ninth letter of Text 2 would be (F−X) which is I (we could also obtain this by looking up Table 7.3 (the decipher table above) at the point where row F (the row of the cipher letter) intersects column X (the column of the assumed plaintext letter in Text 1). The first 15 letters of the cipher and the partial texts then read

```
Cipher    FLIQTNYQFKVACEH
Text1     THE LAST..
Text2     MEETING I.
```

The letter following the I in Text 2 is very likely to be N or S and, in either case, the letter after that is probably X. The corresponding letters in Text 1 would then be either (K−N) and (V−X) or (K−S) and (V−X) that is either XY or SY. The latter is much more likely since the former implies a double space. We tentatively accept SX in Text 2 and the resulting SY in Text 1, and our partial decrypt now reads

```
Cipher    FLIQTNYQFKVACEH
Text1     THE LAST SY..
Text2     MEETING IS ..
```

Our next task is to discover what letter follows SY in Text 1. There are not many candidates and B, L, M, N and S are the most common. Since the cipher letter in position 12 is A the corresponding letter in Text 2 in these five cases is (A−B), (A−L), (A−M), (A−N) or (A−S) that is, Z, P, O, N or I. Of these only Z looks unlikely so we turn our attention elsewhere, to see if we can find some additional clues. Going back to the five possible occurrences of THE we see that we have confirmed the first (at position 1) and ruled out the second (at position 10) so we look at the third (at position 15) which would give us THEX in one text (we can't decide immediately in which of the two texts it occurs) and ONYX in the other. If THEX occurs at position 15 then X ought to be found at position 14 and since the cipher letter there is E the letter preceding ONYX would then be (E−X) which is H. The third word in Text 1 would then be

```
SY..HONY
```

which strongly suggests that it is SYMPHONY. If this is so then since the 12th and 13th cipher letters are A and C, the corresponding letters of Text 2 are (A−M) and (C−P), i.e. O and N, which looks promising, and our decrypt now reads

```
Cipher    FLIQTNYQFKVACEHUCUAC
Text1     THE LAST SYMPHONY ..
Text2     MEETING IS ON THE ..
```

We now turn our attention to the two remaining places where it looks as if THEX might be in one of the texts, namely at positions 39 and 41. These are obviously incompatible, since they overlap, so at most one of them is right. There is a bit more evidence available since if THEX is present it should be preceded by the separator, X. We accordingly decrypt X at positions 38 and 40 where the cipher letters are X and L which give A and O as the corresponding plaintext letters and so yield the possible plaintext pentagraphs

```
ABEEO    at position 38
OPENH    at position 40
```

The first of these doesn't look likely; the second looks better and since a meeting is being mentioned it is quite possible that a place is named. Even without collateral information, since COPENHAGEN fits the pentagraph it is worth trying. We therefore try C at position 39 and A, G, E, N and X in positions 45, 46, 47, 48 and 49 which produces the following texts for positions 39 to 49:

```
Cipher   ULILREJATVQ
Text1    S THE JUPIT
Text2    COPENHAGEN
```

This is very convincing and gives us an extra piece of useful information, for The Jupiter is the name of Mozart's last symphony and we might therefore expect his name to occur in Text 1, somewhere between positions 19 and 38. Furthermore, the word COPENHAGEN should be preceded by a separator and so, putting in the cipher letter, X, and the text letter, which is also X in this case, we obtain A at position 38 of Text 1 which therefore reads

```
Text1    .AS THE JUPITE
```

Since we are dealing with Mozart's last symphony the letter in position 37 might well be W and that at 36 would then be X. Substituting these into the cipher we obtain, for positions 36 to 49,

```
Cipher   FJXULILREJATVQ
Text1    WAS THE JUPIT
Text2    IN COPENHAGEN
```

There is a 50th cipher letter, B, which ought to give E in TEXT1 and 'space' in TEXT2, and this is the case, adding further confirmation, if that were necessary, to the decrypts obtained so far. The word IN in Text 2 should be preceded by a space, so we decrypt X at that point, position 35, where the cipher letter is Q, which produces T in Text 1. The full situation, then, at this point is that we have decrypted positions 1 to 18 and positions 35 to 50, two-thirds of the total, and the cipher and texts are

```
FLIQTNYQFKVACEHUCUACMOXRGEYYQJBNOEQFJXULILREJATVQB
THE LAST SYMPHONY                 T WAS THE JUPITE
MEETING IS ON THE                  IN COPENHAGEN
```

In Text 1 we are looking for the word MOZART and the T at position 35 might be the last letter of it, so we try XMOZAR at positions 29 to 34 which give TXNOON in Text 2. The T is likely to be preceded by XA which yields BY in TEXT1 in positions 27 and 28 and our texts now read

```
FLIQTNYQFKVACEHUCUACMOXRGEYYQJBNOEQFJXULILREJATVQB
THE LAST SYMPHONY         BY MOZART WAS THE JUPITE
MEETING IS ON THE          AT NOON IN COPENHAGEN
```

The word BY should be preceded by X which gives (E−X) = H in Text 2 and since we would expect a date for this meeting we can reasonably try T as the letter before H which gives (G−T) = N in position 25 in Text 1. There are now only 6 letters to be deciphered and in TEXT2 these are almost sure to represent a number, probably a date, which in its ordinal form ends in TH. ELEVEN, with 6 letters, is a good candidate and subtracting these 6 letters from the corresponding cipher letters, ACMOXR, produces WRITTE for Text 1. The decryption is now complete and reads

```
FLIQTNYQFKVACEHUCUACMOXRGEYYQJBNOEQFJXULILREJATVQB
THE LAST SYMPHONY WRITTEN BY MOZART WAS THE JUPITE
MEETING IS ON THE ELEVENTH AT NOON IN COPENHAGEN
```

The cryptanalyst has not only decrypted the message he has also discovered that the book being used as the key is probably a book either about music or about Mozart, and this may be useful in decrypting later messages.

This example, short and simple though it is, illustrates how the cryptanalyst needs a combination of analytical and linguistic skills, general knowledge, imagination and luck in order to achieve success. In addition, do not forget that he would first of all have to realise that the message had been enciphered using a book as key, a fact that is not immediately apparent.

Indicators

These have already been referred to in Chapter 3 but they are sufficiently important to warrant mention again, since provision of the indicator may prove to be the Achilles heel of any cipher system. When a book cipher is used the sender has to let the receiver know at what line on what page of the key book he is beginning the encipherment. If, for example, he begins with line 15 of page 216 he *may* preface the message with the number 15216. This is, however, rather too obvious and he would be more likely to disguise it. There are numerous ways of doing this such as:

(1) transposing the digits according to an agreed rule so that 15216 becomes, say,

65121;

(2) adding an agreed number, such as 59382, digit by digit, thus producing

64598;

(3) converting the digits of the indicator to letters and hiding the resulting five-letter group at an agreed place within the cipher text of the message; thus 15216 becomes

BFCBG;

(4) a combination of any of these.

Provision of an indicator is not, of course, unique to book ciphers; it is an essential part of many cipher systems, but the basic principle is the same: the sender must somehow communicate it to the receiver in such a way as to make it as difficult as possible for a cryptanalyst to find it. The cryptanalyst, conversely, will give high priority to identifying the location of the indicator and discovering its method of encryption.

Disastrous errors in using a book cipher

A significant risk for the sender of a book cipher message is that if he makes a mistake and has to re-send the message he may provide the cryptanalyst with a relatively simple route into decryption. Such a crucial mistake can occur if the sender leaves out a letter of the text of the book key, for example:

Example 7.5 (Spy makes a mistake and gives the game away)
Two messages from the same person and with identical indicators were sent within hours of each other; the cipher texts were

ZECBH MOPJO IIUXJ ELFDR WRSJX CQ.
ZECSS HLIEL RVBCM CUAKA OLPBP PPP

We shall begin the decryption on the assumption that, since the first message has one letter fewer than the second, the sender left out a letter of the book key in the first message and then re-enciphered the text correctly in the second message.

Start of solution
Since the first three letters of the cipher text are the same in both cases we assume that it was the fourth letter of the book key that was not used in the first message. If we were to try all 26 possibilities for this letter in the second cipher message we would obtain 26 possibilities for the fourth letter of the message and, in each case, this would then give us the fifth letter of the book key from the first cipher message. With the fifth letter of the book key we would then recover the fifth letter of the message from the second cipher message, which would then lead us to the sixth letter of the book key from the first cipher message; and so on. We are thus enabled to unravel both the book key and the message from the fourth letter onwards. Of course we initially have to try all 26 possibilities for the fourth key letter but we would quickly be able to see which was the right one when the book key and message texts began to appear. To save time and space we will simply look at what happens when we choose the right letter, which is F, as the fourth letter of the key. We shall refer to the cipher texts as CT1 and CT2; CT2 is the correct text and CT1 has the error. The plaintext letters can be worked out directly, as shown below, or by using Table 7.3.

From CT2: (cipher – key) at position 4 = plaintext letter at position 4; i.e.

S – F = N.

From CT1: (cipher – plain) at position 4 = key letter at position 5; i.e.

B – N = O.

From CT2: (cipher – key) at position 5 = plaintext letter at position 5; i.e.

`S−O=E.`

From CT1: (cipher – plain) at position 5 = key letter at position 6; i.e.

`H−E=D.`

From CT2: (cipher – key) at position 6 = plaintext letter at position 6; i.e.

`H−D=E.`

From CT1: (cipher – plain) at position 6 = key letter at position 7; i.e.

`M−E=I.`

From CT2: (cipher – key) at position 7 = plaintext letter at position 7; i.e.

`L−I=D.`

From CT1: (cipher – plain) at position 7 = key letter at position 8; i.e.

`O−D=L.`

From CT2: (cipher – key) at position 8 = plaintext letter at position 8; i.e.

`I−L=X.`

Looking at the recovered key and message texts so far we have

Key: ... F O D I L
Message ... N E E D X

This looks quite promising, so:

Problem 7.3
Complete the solution which has been started above.

'GARBO''s ciphers

The 'double agent' Jean Pujol (codename GARBO), a Spaniard, used both secret inks and cipher systems during his time in England (1942–5). The inks and ciphers were provided by his German 'control' in Madrid, who was of course unaware that GARBO was working for the British. In 1942 and 1943 GARBO used a simple substitution system based upon 5 alphabets, a very weak system which any competent cryptanalyst would quickly solve. Perhaps realising this, in 1944 he was told to use a different 5-alphabet simple substitution and follow it with double transposition, making the cipher more secure but still vulnerable if enough messages were sent (as they were). Since the British knew the cipher system and were composing the messages with GARBO the cryptanalysts had nothing to do anyway.

GARBO's first cipher

The message was written out as a series of 5-letter groups which were enciphered using 5 substitution alphabets in succession. The first letter of each group was replaced using the first alphabet, the second letter in each group used the second alphabet and so on. This weak system was made even weaker by the fact that the first substitution alphabet was of the Julius Caesar type, each letter being moved 6 places forward, and the other alphabets were not much better. Had GARBO really been working for the Germans his chances of remaining at liberty would have been very poor.

GARBO's second cipher

This involved a 5-alphabet substitution followed by double transposition. The substitution alphabets were now based upon a 5×5 square. This in itself was even weaker than using 5 independent alphabets, but the double transposition significantly improved the security. The 5×5 substitution square was as shown in Table 7.5.

Table 7.5

L	A	C	O	N
F	I	Z	E	G
B	R	T	D	H
J	M	P	Q	S
U	V	W	X	Y

The letter K does not appear in the square; GARBO's messages were normally in Spanish and K would hardly ever be needed; if it did occur, as in York or Kidderminster, it was left as K.

The message was again written as a series of 5-letter groups and the substitution part of the encipherment was carried out as follows:

(1) the first letter of each group was replaced by the letter *above it* in the box;
(2) the second letter of each group was replaced by the letter *to its right* in the box;
(3) the third letter in each group was replaced by the letter *below it* in the box;
(4) the fourth letter of each group was replaced by the letter *to its left* in the box;
(5) the fifth letter of each group *was left unchanged.*

As usual the bottom row of the box was considered as 'above' the first if necessary, the leftmost column as to the right of the rightmost etc. Thus the letter T would encipher at 5 consecutive positions as ZDPRT and the letter N as YLGON whilst a text beginning

STRON GXIND ICATI ONSXT

would after this first stage appear as

HDMCN NYROD AOIRI XLYWT.

A 31-long transposition was now applied. The 31 numbers of the columns were in a fixed but 'random' order. The cipher text was written into the transposition box in rows with the first letter of the text being placed *under the column number corresponding to the day of the month.* The first and last rows would usually contain fewer than 31 letters but the intervening rows would be 'full'. This text was now read out *column by column* starting with the column numbered 1 and proceeding with the columns in numbered order. This transposed cipher text was then written into the same 31-long transposition box with the first letter being placed *under the column number corresponding to the month.* The text was again read out *column by column* starting with the column numbered 1 as before. This doubly transposed text was now written out in 5-letter groups and transmitted.

To illustrate the method of encipherment by GARBO's second system we modify it to use a 12-long transposition and use the month of the transmission to determine the starting column at the first stage and the day of the week (Sunday = 1) to determine the starting column at the second stage.

Example 7.6
Encipher the message

AGENTXFOURXREPORTSXTHATXCONVOYXLEFTXGLASGOWXTODAYX

using GARBO's second cipher with substitution box as above and the 12-long transposition

6 1 10 4 8 11 3 7 12 2 9 5

The message date is Tuesday 18th of May.

Encipherment
We write the message out in 5-letter groups:

AGENT XFOUR XREPO RTSXT HATXC ONVOY XLEFT
GLASG OWXTO DAYXX

(an extra X has been added at the end to complete the last group). The substitution box is as in Table 7.5. The substituted text, using the rules above, is therefore

VFDOT QIEYR QTDMO IDYWT GCPWC XLACY QADGT
NAIQG XXORO ECNWX

The transposition key is
6 1 10 4 8 11 3 7 12 2 9 5

Table 7.6

6	1	10	4	8	11	3	7	12	2	9	5
											V
F	D	O	T	Q	I	E	Y	R	Q	T	D
M	O	I	D	Y	W	T	G	C	P	W	C
X	L	A	C	Y	Q	A	D	G	T	N	A
I	Q	G	X	X	O	R	O	E	C	N	W
X											

and since the month of the transmission is May we must begin entering the text under the column headed 5: see Table 7.6. The text is now taken out column by column starting with the column numbered 1:

DOLQQ PTCET ARTDC XVDCA WFMXI XYGDO QYYXT
WNNOI AGIWQ ORCGE.

Table 7.7

6	1	10	4	8	11	3	7	12	2	9	5
						D	O	L	Q	Q	P
T	C	E	T	A	R	T	D	C	X	V	D
C	A	W	F	M	X	I	X	Y	G	D	O
Q	Y	Y	X	T	W	N	N	O	I	A	G
I	W	Q	O	R	C	G	E				

We now use the transposition box again. Since the *day* of transmission is Tuesday (the *date* is irrelevant in this simplified form of GARBO's cipher) we begin writing the text under the column headed 3: see Table 7.7. Finally, we take the text out column by column, starting with the column numbered 1, and write it out in five-letter groups ready for transmission:

```
CAYWQ XGIDT INGTF XOPDO GTCQI ODXNE AMTRQ
VDAEW YQRXW CLCYO.
```

Decipherment in this system is a tedious process in which it is easy to make mistakes. To begin the decipherment the receiver has to work out, from the day of the week and the length of the message, which columns of the transposition box will have an extra letter and which they are. In the example above since the message contains 50 characters there will be 10 columns of 4 letters and 2 columns of 5 letters. Since the day of the week is Tuesday (= 3) the columns headed 3 and 7 will be the 'long' columns and the others will be 'short'. The same analysis will have to be used when the transposition is used again; in this case the month is May (= 5) so the long columns will be those headed 5 and 6 (since the column headed 6 happens to follow the column headed '5').

For more details of GARBO's cipher systems see [7.1].

One-time pad

The basic weakness of the book cipher as used above is that both the message and the key were in English and by dragging cribs based upon common English words which might occur in either we were able to recover both. Had the key not been in English, decryption would certainly have been more difficult for the cryptanalyst but the messages would nevertheless be read eventually once he had discovered that this was the case, since cribs from the other language could also be used. If, on the other hand, the key was not based upon a natural language but was simply a 'random' string of letters, taken from a page of 'random letters' *which is destroyed after use so that it can never be used again*, then we have what is known as a *one-time pad* and the resulting cipher cannot be solved. This may seem to be a very bold assertion, but it is a mathematical fact [M7].

Since one-time pads provide total security why are they not used for all encipherments? Basically because a different pad has to be provided for every pair of people who need to communicate, each of whom has one of the only two copies produced, and although this is feasible for a few hundred pairs, such as ambassadors communicating with their governments, it is out of the question for large numbers of military units in wartime. It must also be realised that the situation changes dramatically if a 'one-time' pad is used more than once. We then have a 'depth' of two, or more, cipher messages enciphered with the same additive key. By subtracting one text from another the key is eliminated and the resulting *differenced text* is now the difference of two unenciphered texts and a 'crib-dragging' or similar technique may lead to their decryption. If a one-time pad is used to encipher *code groups*, rather than natural language, the cryptanalyst's task is much harder since he must have some knowledge of the code itself in order to use the crib-dragging attack, but the method is essentially the same.

For a variety of reasons, including the provision of one-time pads, cryptographers are very interested in methods by which 'random' letters (or numbers) can be generated. We look at some of the methods of doing this in the next chapter.

8

Producing random numbers and letters

Random sequences

Suppose that we have a long sequence of 0s and 1s, that is a long *binary sequence*. What do we mean when we say that the sequence is 'random'? As an obvious first criterion it seems reasonable to expect that it will contain 'about as many 0s as 1s'; but what do we mean by 'about'?

If the sequence is exactly 1000 digits in length we would not necessarily expect it to contain exactly 500 0s and 500 1s, but if it contained, say, 700 0s and 300 1s we would surely think that it was not a random sequence. Somewhere between these two extremes would mark the limit of acceptability of what *we* would be prepared to accept as random: 530 0s and 470 1s for example; but another person might set different limits. Suppose, however, that the sequence did in fact consist of 500 0s followed by 500 1s. Since there are exactly 500 of each digit can we consider the sequence to be random? Clearly not, since in a random sequence we would expect the four two-digit numbers, 00, 01, 10 and 11, each to occur 'about 250' times but in this sequence 00 and 11 both occur 499 times, 01 occurs only once and 10 doesn't occur at all. Even if the sequence passes *this* test we could then ask whether the eight three-digit numbers 000, 001, 010, 011, 100, 101, 110 and 111 each occur 'about 125' times, and so on. An endless variety of requirements of such types can be proposed, and there is an extensive mathematical literature on tests that can be applied to a sequence to see if it might reasonably be said to be 'random'. Conversely, there is also an extensive literature describing methods for producing sequences which, whilst they are not strictly 'random', satisfy various randomness tests and so are considered to be sufficiently unpredictable to be useful in certain situations. Without going into the mathematical criteria for deciding if a sequence is random a suitable definition for our purposes is

Definition 8.1
A binary sequence is considered to be random if, no matter how many digits we have seen, the probability that the next digit will be 0 is 0.5.

This is the situation that should apply if one spins an 'unbiased' coin many times: no matter what has already happened the probability that it will come down 'heads' next time should be 0.5, or in terms of odds, 'evens'.

There is nothing special about *binary* sequences; our definition of randomness can be applied with only slight modification to sequences of decimal digits or letters.

Definition 8.2
A sequence of decimal digits is considered to be random if, no matter how many digits we have seen, the probability that the next digit will have a particular value is 0.1.

Definition 8.3
A sequence of letters from the English alphabet is considered to be random if, no matter how many letters we have seen, the probability that the next letter will be a particular one is 1/26.

Producing random sequences

A truly random sequence can only be generated by a truly random process and so, in particular, cannot be generated by any mathematical formula, for knowledge of the formula and sufficient initial values (i.e. of numbers already generated by the formula) would enable someone to predict the next value with certainty. There are, however, formulae which can produce a long sequence of numbers which satisfy many randomness criteria before they start to repeat; such sequences are called 'pseudo-random' and we describe some of these below, but first we look at some ways of generating truly random sequences.

Coin spinning

If we spin a 'fair' coin many times and write down '1' each time it comes up 'heads' and '0' each time it comes up 'tails' we ought to get a random binary sequence. In practice, perhaps because of some regularity about the spin, the sequence may be biased. It is, in addition, a very slow way of producing a random sequence that would presumably only be used if no other method were available. It is said that a prisoner of war carried out such a procedure for many thousand spins, to keep himself occupied, and analysed the resulting sequence with a variety of tests.

Throwing dice

A less laborious procedure can be based on throwing two dice. The dice must be distinguishable one from the other; let us assume that one is coloured red and the other is coloured blue. Throw both dice and compute the number

$$6 \times (\text{number on the red die}) + (\text{number on the blue die}) - 7$$

then

(i) reject the number if it exceeds 29,
(ii) write down the remainder when the number is divided by 10.

The resultant sequence of decimal digits should be random.

The rather odd-looking rules are necessary because the faces of the dice are numbered 1 to 6 and not 0 to 5 and because there are 36 combinations which can be produced. Consequently values from 0 to 35 inclusive for the number can be produced and so we must reject any number above 29, in order to ensure that all the digits from 0 to 9 have an equal chance of appearing.

More than two dice may be used and then more than one random digit can be generated at each throw. With four dice, for example, there are 1296 possible outcomes and if we colour the dice red, blue, green and white and compute the number

$$216 \times \text{red} + 36 \times \text{blue} + 6 \times \text{green} + \text{white} - 259$$

and reject any number above 999 we can take the three-digit number so obtained as the next three digits of the random decimal sequence.

There are many possible variations on this type of approach; for example, the two dice could be replaced by a roulette wheel, which has 37 sectors numbered 0 to 36. Sectors 30 to 36 would be ignored and the second digit of the 'winning' sector would provide the next random decimal digit. This is obviously rather wasteful and a more efficient use in this case would be to ignore sectors 32 to 36 and convert the other numbers, 0 to 31 inclusive, into binary, thus providing five binary digits each time. Binary digits are commonly known as *bits* and are frequently referred to in that way. Binary keys are frequently used in cryptography. Not only have they the great merit that non-carrying addition (mod 2) is particularly simple and identical to non-carrying *subtraction* (mod 2), which makes encipherment and decipherment the same, but also (mod 2) arithmetic is very easy to simulate electronically, and so is particularly suitable both for cipher machines and for simulators on computers.

Lottery type draws

The system used to draw lottery (or bingo) numbers could be used provided it was modified so that a number which has been drawn is immediately returned to the pool. Thus 100 balls numbered 00 to 99 are spun in a barrel and selected one by one, each number selected is noted and provides two decimal digits for the table of random numbers. The selected ball *must* be put back in the barrel for otherwise it couldn't be drawn again and each page of 100 two-digit decimal numbers would contain each number once, and only once, and so would not be random. A typical page of 100 two-digit random numbers would be expected to contain some numbers three, or even four, times whilst between 30 and 40 numbers might not occur at all. (For an explanation see M8.)

Cosmic rays

Cosmic rays are produced when particles from the Sun enter the Earth's atmosphere and generate cascades of other particles by collisions and so provide a 'natural' source of (presumably) random events. If we were to install ten detectors, such as Geiger counters, numbered 0 to 9, in a room and record the order in which the detectors 'fire' we would obtain a genuinely unpredictable decimal sequence. Care would have to be taken that when a detector has 'fired' no other event is recorded until that detector has had time to 'recover', for otherwise there is likely to be a deficiency of 'doublets' such as 00, 11 etc. in the resultant sequence.

Amplifier noise

Noise in electrical circuits is usually regarded as a problem, but it can also be turned to good use in cryptography. The noise can be converted into a signal which is used to switch a *gate* on or off, and this in turn is then interpreted as a 0 or a 1. If the circuits are carefully adjusted the binary stream so produced should be effectively random. If there is some residual bias, in that the probabilities of 0 and 1 occurring are slightly different from 0.5, the (mod 2) sum of two or more such streams will reduce it considerably. Two *unrelated* streams each of which has a bias of 0.51 to 0.49 in favour of 0, for example, will combine to produce a stream with a bias of only 0.5002 to 0.4998 (M9).

Pseudo-random sequences

We have already encountered the Fibonacci sequence in Chapter 6. This is an infinitely long sequence of integers generated by the simple rule that each number in the sequence is the sum of the two previous numbers. The sequence is traditionally started by taking the first two numbers as 0 and 1. The Fibonacci sequence unfortunately has many arithmetic properties, as was mentioned before, and so is quite unsuitable as a source of pseudo-random numbers. Suppose, however, that we modify the rule to, say, that each number is the sum of *twice* the previous number plus the number before that, would we get a better sequence for our purposes? If we begin with 0 and 1 as the first two terms, the first 10 terms of the sequence are

$$0, 1, 2, 5, 12, 29, 70, 169, 408, 985.$$

It will be noted that the terms are even and odd alternately and this, by itself, is sufficient to rule them out as a source of pseudo-random numbers. Of course we needn't begin with 0 and 1 as the first two terms, we could start with any two numbers, but the flaw is fundamental and no sequence generated in this way would be satisfactory. The sequence, as might be expected after seeing the many features of the Fibonacci sequence, has many mathematical properties; for example, every third term is divisible by 5 and the ratio of consecutive terms rapidly approaches the fixed value

$$2.414\,213\,56\ldots$$

which is

$$(1 + \sqrt{2}).$$

(For more detail see M10).

Linear recurrences

The sequences looked at above are examples of sequences generated by means of what are known as *linear recurrences*. Since each new term involved adding together multiples of the *two* preceding terms they are more specifically known as *linear recurrences of order* 2. More generally, a *linear recurrence of order k* is one in which each new term is the sum of multiples of the k preceding terms. So, for example, if we let U_n denote the nth term of a sequence then

$$U_n = U_{(n-1)} + 2U_{(n-2)} - U_{(n-3)}$$

is a *linear recurrence of order* 3 and

$$U_n = U_{(n-3)} + U_{(n-5)}$$

is a *linear recurrence of order* 5. The fact that in the second case some of the preceding terms are not involved doesn't matter; five preceding terms are required in order to find the next term but three of the terms, $U_{(n-1)}$, $U_{(n-2)}$ and $U_{(n-4)}$, have multipliers of 0. Had the term $U_{(n-5)}$ not been present however the recurrence would not have been of order 5. The multipliers, for our purposes, are always integers but may be positive, negative or zero. It is assumed that in a linear recurrence of order k the term $U_{(n-k)}$ *is* present, with either a positive or a negative multiplier, but not zero.

The terms of linear recurrences usually grow very rapidly and although they often have interesting arithmetical properties they are only suitable for cryptographic purposes when the terms themselves are replaced by their values (mod 2), that is the terms are replaced by 0 if they are even and by 1 if they are odd, thus producing a *binary sequence.* Calculation of the terms of a linear recurrence (mod 2) is particularly easy, there is no need to compute the actual value of the terms and then replace them by 0 or 1. Each term is simply replaced by 0 or 1 *as soon as it is calculated*; we then only have to add up a number of 0s and 1s which is a lot easier than adding increasingly large integers. The resulting binary sequence is identical to the one which would be obtained by computing each term exactly and then replacing it by 0 or 1. So, for example, the linear recurrence of order 2

$$U_n = 3U_{(n-1)} - 2U_{(n-2)}$$

with the initial values $U_0 = 0$, $U_1 = 1$ continues

$$0, 1, 3, 7, 15, 31, 63, 127, 255, 511, \ldots..$$

If we replace each term by its remainder (mod 2) as soon as it is calculated the binary equivalent is

$$0, 1, 1, 1, 1, 1, 1, 1, 1, 1, \ldots$$

which is clearly correct since, obviously,

$$U_n = 2^n - 1$$

and all the terms after U_0 are odd.

This particular sequence is clearly of no use to a cryptographer since it is exceptionally non-random. Is it possible however that some binary linear sequences might be suitable, and how would they be used? We look first at the practical problem of how a binary stream of key could be used for encipherment.

Using a binary stream of key for encipherment

The cryptographer would first have to convert the text of the message from an alphabetic/numeric to binary form. In the early days of computers five or six bits were used to represent the most important characters. Since these provided for only 32 or 64 characters respectively, which imposed limitations on the character set that could be used, they were eventually replaced by an eight-bit representation, which became known as a *byte*, allowing 256 characters, sufficient to include not only lower and upper case letters, numbers and punctuation but also numerous other symbols such as brackets of various kinds and accents. Today an eight-bit representation is standard, for example,

A = 65 = 01000001,
B = 66 = 01000010

etc. and

a = 97 = 01100001,
b = 98 = 01100010,

whilst

\$ = 36 = 00100100

and

ê = 136 = 10001000.

Having converted the characters of text to eight-bit bytes they would then normally be added 'linearly' (mod 2), that is bit by bit without 'carrying', to the binary key to produce the cipher. So, for example, if the message letter is E and the corresponding key letter is $,

E = 01000101,
$ = 00100100.

Adding (mod 2)

01100001 = 97 = a,

i.e. the cipher letter is a.

There are alternatives to this 'linear' or bit-by-bit addition; a very important and secure system, the Data Encryption Standard (DES) described in Chapter 13, treats some of the bits in a 'non-linear' way.

Binary linear sequences as key generators

When we generate a binary sequence by using a linear recurrence of order k we produce a sequence of 0s and 1s. Could this sequence go on indefinitely without repeating? The answer is 'No' because each new term depends only on the values of the previous k terms and since each of these is 0 or 1 there are only 2^k different possibilities for them. It follows that after 2^k terms, at most, some set of k consecutive binary terms *must* recur. Thus the Fibonacci sequence (mod 2) is

$$0, 1, 1, 0, 1, 1, 0, 1, 1, 0, \ldots$$

and we see that the binary sequence consists simply of the triplet 011 repeated indefinitely. Since the Fibonacci sequence is generated by a linear recurrence of order 2 we have $k = 2$ in this case, and so we know that in binary form the sequence *must repeat* after at most $2^2 = 4$ terms. In fact it repeats after 3 and this is, in reality, the maximum that it can be, because one of the 4 possible pairs of consecutive binary terms is 00 and such a pair will generate 0s for ever. Conversely, no other binary sequence can contain 00 and so the maximum possible number of binary terms that we can have before the sequence starts to repeat is 3, not 4. For the same reason the maximum number of terms before a binary linear recurrence of order k begins to repeat is $2^k - 1$, not 2^k. The binary Fibonacci sequence, modest though it is, therefore has maximum period.

Obviously no binary sequence with a maximum period of 3 is of interest to cryptographers but what about binary sequences of higher order?

To take an easily verifiable case first: a sequence of order 4 might have a period of 15, which is 2^4-1. If we ignore the trivial case where all four multipliers are 0 there are 15 possible binary linear sequences of order 4. Do any of these produce a binary sequence of maximum period 15? Just 2 of them do; they are

$$U_n = U_{(n-3)} + U_{(n-4)}$$

and

$$U_n = U_{(n-1)} + U_{(n-4)}.$$

Example 8.1

Verify that the binary linear recurrence of the fourth order

$$U_n = U_{(n-3)} + U_{(n-4)}$$

generates a sequence of maximum period 15.

Verification

We start with $U_0 = U_1 = U_2 = U_3 = 1$ and generate each new term by adding together the terms three and four places earlier in the sequence and putting 0 or 1 according to whether the sum is even or odd. The sequence is then found to be

1, 1, 1, 1, 0, 0, 0, 1, 0, 0, 1, 1, 0, 1, 0, <u>1, 1, 1, 1</u>...

and we see that the sequence begins to repeat from U_{15} onwards, but not before.

(Note that if a binary sequence generated by a linear recurrence of order k has maximum period it must contain all possible 2^k binary sequences of length k except the one consisting of all 0s. We may therefore take *any* initial values except 00...00 and the period will be seen to be maximal in every case. If the period is not maximal different starting values may produce different sequences.)

Problem 8.1

Verify that the binary recurrence

$$U_n = U_{(n-1)} + U_{(n-4)}$$

also generates a sequence of period 15 but that the recurrence

$$U_n = U_{(n-1)} + U_{(n-2)} + U_{(n-3)} + U_{(n-4)}$$

does not.

What about sequences of higher order? A sequence of order 12, for example, might have a period as long as $2^{12}-1$, which is 4095. Does any linear recurrence of order 12 generate such a maximal binary sequence? As a result of some elegant and advanced mathematics a formula is known which tells us exactly how many binary linear recurrences of order k will produce a sequence of maximum period. In the case of recurrences of order 12 the formula tells us that 144 binary linear recurrences of order 12 will produce binary sequences of period 4095. (The fact that $144 = 12 \times 12$ is a fluke!) The remaining 3952 will fail to do so. The mathematical analysis does not lead directly to these 144, which must be found by a process somewhat akin to finding prime numbers. Alternatively, using a computer, the 4095 possible recurrences can be tested and any which repeat before 4095 terms have been generated can be rejected. When this is done the first successful sequence found is

$$U_n = U_{(n-6)} + U_{(n-8)} + U_{(n-11)} + U_{(n-12)}.$$

It is the 'first' in the sense that, writing 0 and 1 for the multipliers of the 12 terms on the right-hand side of the linear recurrence, this sequence has the 12-bit representation

000001010011

which, interpreted as an integer written in binary form, is

$$64+16+2+1=83.$$

No sequence with such an integer representation below 83 produces the maximum period of 4095. For some of the mathematics behind all this see M11.

By choosing a sufficiently high order and finding a linear recurrence that gives the maximum period we can produce a key stream that would seem to provide a pseudo-random binary stream which we could use as a key for encryption. It can be shown, for example, that 356 960 linear recurrences of order 23 will generate maximal key streams, which are more than 8 000 000 long (M11). Since the initial starting values also provide over 8 000 000 possibilities one might think that such a key would present a formidable problem to the cryptanalyst and, initially, it would. Unfortunately for the cryptographers key which has been generated in this way has a fatal flaw: given a fairly modest amount of the key the linear recurrence by which it has been generated and the initial values can be recovered This is a consequence of the following.

Cryptanalysis of a linear recurrence

Given 2k consecutive bits of key generated by a binary linear recurrence of order k a system of k linear equations in the k multipliers of the terms of the recurrence can be set up and solved.

If the cryptanalyst has reason to believe that key generated by a linear recurrence is involved he would proceed as follows:

(i) obtain a stretch of key from a message to which the solution is known; not many characters will be required and may be available from standardised beginnings to messages;
(ii) assume a value for k, the order of the linear recurrence;
(iii) using $2k$ consecutive bits of the key set up k linear equations in the k unknown multipliers in the recurrence; if the recurrence *is* of order k a solution will appear in which all the multipliers are integers (and are to be interpreted (mod 2), i.e. as odd or even); it is possible that there is more than one solution of order k or, alternatively, that there is *no* such solution; in the latter case a different value of k should be tried.

Thus, with characters represented as 8-bit bytes, and a key stream generated by a binary linear recurrence of order 23, only 46 bits of key would be required in order to solve the system. Since 6 characters of the message would produce 48 bits of key the system could be solved easily if messages tended to begin `TO THE`!

For worked examples of these cases see M12.

Improving the security of binary keys

It is clear that binary keys generated by linear recurrences are too easy to solve to be useful from a cryptographic point of view, but is there any way in which their security can be improved? Since their weakness lies in the fact that in a recurrence of order k each bit is a fixed linear combination of the k bits which precede it even using a recurrence of high order, such as taking $k=103$, will not provide sufficient security, for only a moderate stretch of key (26 letters when $k=103$) would be required to recover the system. In addition, use of a recurrence of high order to generate the key manually (as a spy might have to do) would be a tedious task and prone to error. This is a pity, since a key generated from such a recurrence could have a very long period, more than 10^{30} when $k=103$. A long period is highly desirable but can we obtain this without using a high order recurrence and strengthen the security at the same time? In fact we can: by combining the keys of two or more linear recurrences, as the following simple example shows.

Example 8.2
Use the (mod 2) sum of the keys generated by the two linear recurrences

$$U_n = U_{(n-1)} + U_{(n-2)},\ U_0 = U_1 = 1$$

and

$$U_n = U_{(n-1)} + U_{(n-3)},\ U_0 = U_1 = U_2 = 1$$

to produce a new key. Verify that this has a period of 21.

Verification
The first recurrence, as we have seen, has a period of length 3 and produces the keystream

110110110110...

The second recurrence has a period of length 7 and produces the key stream

111010011101001110100....

Writing both of these out one under the other and adding the bits (mod 2) we have

	1 1 0 1 1 0 1 1 0 1 1 0 1 1 0 1 1 0 1 1 0 1 1 0 1 1 0.....
	1 1 1 0 1 0 0 1 1 1 0 1 0 0 1 1 1 0 1 0 0 1 1 1 0 1 0.....
Adding (mod 2)	0 0 1 1 0 0 1 0 1 0 1 1 1 1 1 0 0 0 0 1 0 0 0 1 1 0 0.....

and we see that the key repeats after 21 places, but not before. Since the first key has period 3 and the second key has period 7 the period of the combined key cannot exceed 21, for both keys repeat after 21 places. On the other hand, since 3 and 7 have no common factor, the combined key cannot repeat *after less than 21 places*.

There is no need to restrict ourselves to the use of *two* linear recurrences; we could use three or more. The advantage would be that the more we use the more difficult it would be for a cryptanalyst to solve the system. The disadvantage, if we are working by hand, would be the tedious nature of the key generation and the increased probability of errors. Of course if we have a means of generating the key by either a mechanical or an electronic device the disadvantage disappears. It is not, therefore, surprising that machines have been built which generate long period key streams, both binary (i.e. (mod 2)) and alphabetic (i.e. (mod 26)), by combining the output of several key streams of shorter periods. One such machine, which generated (mod 26) key, was the Hagelin cipher machine which was widely used by several countries in World War II, and another, which generated (mod 2) key, was the Lorenz SZ42 which was one of the cipher machines used by the Germans. These are described in Chapters 10 and 11.

Pseudo-random number generators

It is sometimes necessary to use random numbers in computations, such as those of the 'Monte Carlo' type [8.3], where analytical methods are infeasible, such as problems in particle physics or the dynamics of star clusters. In such cases mathematical methods for producing a stream of pseudo-random numbers are often employed and whilst such numbers are not considered to be suitable for use as keys in cryptographic systems they are of some intrinsic interest and are worth noting.

The mid-square method

Although this method cannot be recommended for use in cryptography it *has* been used in some other applications. If we take an integer at random and square it the last digit cannot be 2, 3, 7 or 8. It would therefore seem that starting with a random number and repeatedly squaring it in the hope of producing a sequence of random numbers is pointless. If, however, after squaring a number we 'throw away' some of its leading and trailing digits the remaining digits might be sufficiently uniformly distributed to be used as a source of pseudo-random numbers. This is the basis of the mid-square method which works as follows:

(1) choose a large 'random' integer, X, of length n digits;
(2) form X^2 and retain only the middle n digits (put a 0 at the front if necessary); use the resulting integer as the new value of X.

The sequence of n-digit integers generated in this way must eventually begin to cycle since there are only 10^n possible values. The cycling may arise from the repetition of a number, not necessarily the starting value, which has appeared previously or by the integer consisting of all 0s being generated, after which only 0s will appear. Care must be taken in using this method for any purpose, as the following example shows.

Example 8.3

(1) Use the mid-square method to obtain four-digit numbers starting with 3317.

(2) Repeat the exercise but start with 2907.

(3) Repeat the exercise but start with 3127.

(A calculator or computer is needed for the second and third cases.)

Solutions

(1) $X = 3317$ so $X^2 = 11\,002\,489$. Removing the first and last pairs of digits gives the next value of $X = 0024$ so $X^2 = 00\,000\,576$ and removing the first and last pairs gives the next value $X = 0005$. Since $X^2 = 00\,000\,025$, the next and all subsequent values of X are 0000.

(2) This is a less extreme case. The values of X begin 2907, 4506, 3040 and seem to be continuing satisfactorily but nevertheless the 42nd value of X turns out to be 0.

(3) In this case cycling occurs but the sequence does not restart from the beginning. Starting with $X = 3127$ we find that the sequence continues 7781, 5439, 5827, ... but from the 38th term onwards we get 6100, 2100, 4100, 8100, 6100, 4100,.... and the same four numbers now repeat indefinitely.

Although mid-squaring is a conveniently simple method it should only be used, if at all, with much larger numbers than in the example.

Linear congruential generators

The most commonly used method generates a sequence of integers in the range 0 to $(M-1)$ by means of a recurrence formula of the type

$$U_n = AU_{(n-1)} + B \pmod{M}$$

where A, B and M are integers. A is called the *multiplier*, B is the *increment* and M is the *modulus*. The process is started off by choosing a value, known as the *seed*, in the range 0 to $(M-1)$ for U_0. Such a recurrence must eventually repeat and the maximum period obviously cannot exceed M, so M should be 'large'. For suitably chosen values of A, B and M a long period *is* attainable. In the best cases the period is maximal and so the choice of the seed is irrelevant, since all but one of the possible values (mod M) occur. The majority of random number generators used in computers are based upon this method with the values of the modulus, increment and multiplier built into the program, and the choice of seed left to the user. To ensure even better results more than one such generator may be used and their outputs combined in some way. Further improvement can be achieved if the numbers are not used in the order in which they are produced, but some form of 'shuffling' is employed to reduce the risk of correlation between consecutive numbers. In this way good, long, 'pseudo-random' sequences may be generated. Among the known 'good' choices for A, B and M are those shown in Table 8.1.

Table 8.1

A	B	M
106	1 283	6 075
171	11 213	53 125
141	28 411	134 456
421	54 773	259 200

At the other extreme, a single generator with badly chosen values of A, B and M may produce key with a very short period. Here is a small scale example to illustrate these situations.

Example 8.4
Use the recurrence

$$U_n = 3U_{(n-1)} + 4 \pmod{17}$$

to generate a sequence of integers starting with (1) $U_0 = 5$, (2) $U_0 = 15$.

Generation
(1) Since $U_0 = 5$, $U_1 = 3 \times 5 + 4 = 19 \equiv 2 \pmod{17}$ etc. The 16-long sequence is

$$5, 2, 10, 0, 4, 16, 1, 7, 8, 11, 3, 13, 9, 14, 12, 6, \underline{5, 2, 10}, \ldots.$$

This generator, though modest, produces the maximum possible cycle, which is 16 in this case. For additional comments see M13.
(2) $U_0 = 15$ gives $U_1 = 3 \times 15 + 4 = 49 \equiv 15 \pmod{17}$, so the period is 1! This explains why the value 15 doesn't occur in the 16-long cycle above.

Problem 8.2
Verify that the mid-square method which uses four-digit numbers starting with $X = 7789$ degenerates into a sequence of four numbers.

Problem 8.3

Starting with $U_0 = 1$ what are the cycle lengths of the recurrences

(1) $U_n = 3U_{(n-1)} + 7 \pmod{19}$,
(2) $U_n = 4U_{(n-1)} + 7 \pmod{19}$?

When pseudo-random number generators of this type are used the values obtained are usually divided by the modulus, M, to give a real number lying between 0 and 1.0. Since the integer values given by the generator are in the range 0 to $(M - 1)$ the values of the real number may include 0.0 but will not include 1.0 but this small restriction is unlikely to be important and, with well-chosen values of A, B and M, the real numbers produced should be uniformly distributed between 0 and 1.

For more information see [8.4] and M13.

9

The Enigma cipher machine

Historical background

In Chapter 2 we looked at simple substitution ciphers and we saw how these can be solved by the use of frequency counts if 'sufficient' cipher text is available. How many letters are always 'sufficient' is a matter for debate, but it is probably true that 200 letters will normally suffice whereas 50 might not. For our purposes let us assume that if only 25 letters of cipher are available then the cipher is safe. Since a limitation of message lengths to no more than 25 letters would be too restrictive we conclude that the use of a simple substitution cipher is impractical. If, however, we use not *one* but *several* different simple substitution alphabets, switching between the alphabets every time we encipher a letter, we can increase the security of the system. As a rough guide: if we use N different alphabets it should be possible to make the cipher safe for single messages of up to $25N$ cipher letters; but this simple rule needs qualification. If the substitution alphabets are related in some way the recovery of any one of them may lead to recovery of the others. On the other hand, in some systems, additional features may ensure that cipher messages of much greater length than $25N$ are secure. In the specific case of Jefferson's cylinder, for example, the sender and receiver could

either agree that the cipher text will be read from the row of letters at a specified distance from the row of plaintext letters (the distance possibly being given by some form of *indicator*),
or have no indicator, and use a different distance each time a row is enciphered.

Whilst the latter procedure involves the recipient in looking at all 25 rows of the cylinder to see which of them makes sense, the security of the system would be considerably enhanced. With the 'fixed distance' system and a cylinder of, say, 40 discs, cipher letters 40 positions apart would come from identical simple substitution alphabets. It follows that a collection of messages containing more than, say, 2000 letters would be vulnerable to attack based upon monograph frequency counts, since all the messages would be 'in depth' and we would have a sample of 50 cipher letters from each alphabet. In the *variable distance* system the messages would not be 'in depth' and thousands more cipher characters might be needed to solve the system; the number needed would obviously depend upon how randomly the variable distances were selected.

Evidently, a system based upon N substitution alphabets has a security level that increases with N but, on the other hand, if the encipherment is to be done by hand the tediousness of using the system, and the possibility of error, would also increase with N. So, as so often happens in life, we have conflicting requirements. In this case we would like to make N large to increase the security, but we would also like to keep N small for ease of use, and we can't do both.

During the 1914–18 war radio began to be used by military units for sending messages to each other and to their headquarters. Radio transmission had the advantage that communication with units at considerable distances from base, including ships and submarines at sea, could be achieved almost immediately, but the disadvantage that the messages could also be intercepted by the enemy. It was therefore essential to encipher such messages in a secure system and cipher systems of some complexity were devised; unfortunately, the more complex the system the greater the burden on the cipher clerks, and the greater the risk of errors with, possibly, disastrous consequences. Some 'user-friendly' but highly secure cipher systems were needed if the conflicting requirements were to be met.

Following the War a number of people in various countries decided that the only way of providing a high level of security without obliging cipher clerks to carry out lengthy, tedious and error-prone processes was to use machines to do the encipherment/decipherment. One such person was Arthur Scherbius, co-founder of a German engineering firm. In the early 1920s Scherbius designed a number of cipher machines, all of which were intended to provide a very large number of substitution alphabets. A different alphabet would automatically be used every time a letter was enciphered, and no substitution alphabet would recur until thousands of letters had been processed. Having decided upon a particular design he constructed the machine and called it Enigma.

The original Enigma

The Enigma which Scherbius constructed and showed at the Universal Postal Union Congress in Vienna in 1923 was based upon the following components:

(1) a 26-letter keyboard for inputting the plaintext message;
(2) 26 lamps which would light up to show the cipher letters;
(3) a power supply (a 3.5 volt battery or equivalent);
(4) three removable wired wheels which could rotate about a common axis;
(5) a fixed wired *reflector*;
(6) a fixed wired *entry wheel*.

The ***keyboard*** was similar to the keyboard on English language typewriters with some minor exceptions, viz: (i) the letters Y and Z were interchanged, so that Z was on the top row and Y was on the bottom row and (ii) the letter P was on the bottom row, not the top. Only upper case letters were used, there were no numerals, nor were there any letters with umlauts, such as Ü. The same arrangement applied to the letters on the lamps.

The ***battery*** was used only to send a current through the wheels and the reflector, and to light up the lamps. It *did not* provide the power to move the wheels, which was done mechanically.

Inside each ***removable wheel*** there were 26 wires which 'randomly' connected 26 contact points on one side of the wheel with 26 contacts on the other side of the wheel. The contact points on one side of the wheel (the left side when looked at from the front of the machine) were flush with the wheel's face, but the contacts on the other side (the 'right' side) jutted out from the face on little springs; this was to provide good contact between a wheel and the one next to it. Similarly, good contact was ensured between the rightmost wheel and the entry wheel and between the leftmost wheel and the reflector. An alphabet 'tyre' ran round the circumference of each wheel and on the left-hand side of each removable wheel a metal ring, the 'notch ring', was attached which had one V-shaped notch in it opposite one of the letters on the tyre. On the right-hand side of these wheels there was a toothed ring with 26 teeth, the *setting ring*, which enabled the cipher operators to turn the wheel to any desired position.

(The word 'randomly' in relation to the wheel wirings needs some qualification but an explanation involves some mathematics, which will

Plate 9.1 One side of an Enigma wheel. The 26 spring contacts are on this side and the smooth-notched setting ring surrounds the wheel itself. The machine identification is M3564 and the wheel is identified as number 2 (in Roman numerals).

be found in M14.) See Plates 9.1 and 9.2 for photographs of both sides of an actual wheel from an Enigma machine.

The ***reflector*** was fixed in position and had 26 contacts on one side only. Inside the reflector 13 wires connected the 26 contacts in pairs and so a current entering one of the contact points of the reflector would exit at another of the contact points. The internal wiring of the reflector was also 'random'. Unlike the three wired wheels, the reflector was permanently fixed in the machine and was only replaced once during the period 1930–45, in 1937.

The ***entry wheel*** provided the connection between the rightmost wheel and the keyboard, and between the rightmost wheel and the lamps. Somewhat surprisingly, the entry wheel was connected to the keyboard letters in normal alphabetic order, rather than in keyboard order. This gave no cryptographic advantage and must have involved some messy internal wiring.

A simplified schematic diagram of the Enigma is shown in Figure 9.1.

The machine was housed in a wooden box. When the cover of the machine was closed only the setting rings on the three movable wheels

Plate 9.2 The other side of an Enigma wheel. The 26 flat contacts are on this side. The alphabet tyre, setting ring and notch ring can be seen. The notch ring has a single notch, visible opposite *M* on the alphabet tyre.

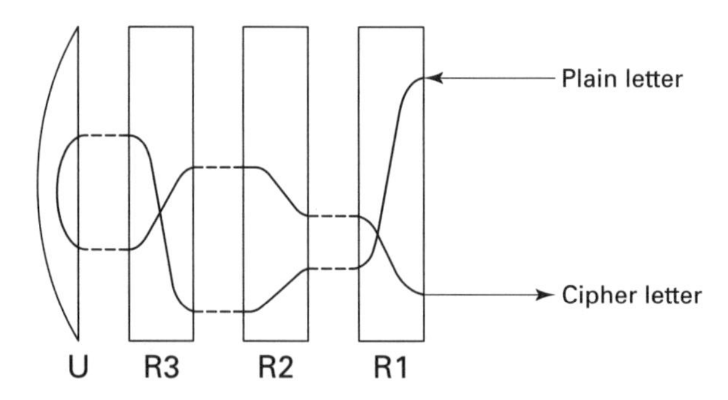

Figure 9.1. The Enigma machine.

protruded, but a letter on the alphabet tyre on each removable wheel was visible through a small 'window' above it. Operators were thus enabled to use the setting rings to turn each wheel to its desired starting position. When the cover of the machine was open the operator could see all the internal wheels and, by using a lever next to the reflector, take the three movable wheels out of the machine, slide them off their common axis,

Plate 9.3 The Enigma with lid closed, ready for use. The current positions of the three wheels are visible through the windows.

and rearrange their order. Since the original Enigma, unlike later versions, had only three wheels available the number of possible wheel orderings was only six. The machine could be carried, but it was quite heavy: about 12 kilograms (nearly 30 pounds).

See Plates 9.3 and 9.4 for photographs of an Enigma with the cover closed and open.

The three removable wired wheels are R1, R2 and R3. The fixed reflector is U (its German name was Umkehrwalze). The entry wheel, battery, keyboard and lamps arc not shown in this simplified diagram. When one of the keyboard letters is pressed a contact is made which causes the current from the battery to pass through R1, R2, and R3. It is then 'turned round' by the reflector, after which it passes through R3, R2 and R1 before lighting up a lamp to show the cipher letter.

The path taken by the current imposes two important features on the cipher:

(1) no letter can encipher to itself;
(2) there is symmetry (or 'reciprocity') of plain–cipher pairs, e.g. if `A` enciphers to `K` then, at the same setting of the wheels, `K` will encipher to `A`.

Plate 9.4 The Enigma with the top and flap open. The three wheels and reflector are now visible as is the plugboard at the front.

Encipherment using wired wheels

Before we can understand how encipherment was carried out in the Enigma we need to see what happens when current from the keyboard passes through a single wired wheel. Since there are 26 contact points on each side of the wheel and the wires connecting the pairs of points on opposite faces are 'randomly placed' the current entering at, say, A will emerge at one of the 26 points on the opposite face. We cannot predict the exit point unless we know the internal wiring of the wheel, but let us suppose that it is Y. If the wheel now turns, the wire that carried the current from A to Y will move one position on each side and will now carry the current from B to Z. Likewise, if B was connected to M and C was connected to A before the wheel turned then C and D will be connected to N and B respectively after the wheel has turned. This is illustrated in Figure 9.2.

When the wheel has turned 26 times the wires will be back in their original positions and A, B and C will again be connected to Y, M and A respectively.

If we know the wiring connections inside a wheel we will know the encipherment of every letter, A, B, C, ..., Z at position 1 of the wheel and we can then work out the encipherment of any letter at any position of the

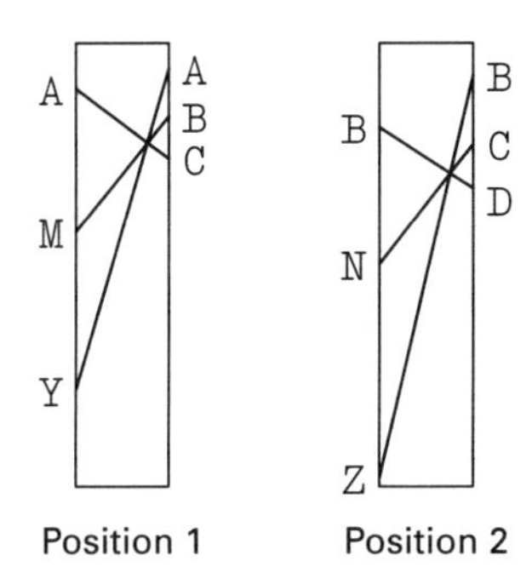

Figure 9.2.

wheel. For example, if we wish to know the encipherment of K at position 6 of the wheel we argue as follows.

The wire that has entry point K at position 6 is the wire that at position 1 had entry point at the letter which is 5 places before K in the alphabet, that is F. If F enciphers to P, say, at position 1 then K will encipher to the letter which is 5 places beyond P in the alphabet at position 6, that is, U. In short:

If F goes to P at position 1 then K will go to U at position 6.

We can fully describe the encipherment property of any wheel by listing the encipherment of each letter at position 1 of the wheel, for we can then work out the encipherment of any letter at position 2, then at position 3, and so on. There is nothing special about position 1, a list of the encipherment alphabet (which is a simple substitution) at *any* position of the wheel will do as well.

So, for example, if we take the encipherment of the first 6 letters at position 1 we can begin to form Table 9.1.

Table 9.1

	Position					
Letter	1	2	3	4	5	6 ...
A	Y	.	.	.	.	.
B	M	Z	.	.	.	.
C	A	N	A	.	.	.
D	T	B	O	B	.	.
E	F	U	C	P	C	.
F	R	G	V	D	Q	D

The 'dots' indicate places where we do not yet have enough information to give the cipher letter. The full 26×26 encipherment table can be filled in when we know the encipherment of all 26 letters at position 1.

Note the important feature of the encipherment table: each diagonal from North-West to South-East constitutes a full alphabet, in normal order, starting from the letter in column 1.

Alternatively, if we know the encipherment of A, or any other letter, at all 26 positions of the wheel we can equally well work out the encipherment of any letter at any position. For example, suppose that we wish to know the encipherment of N at position 11. The wire that has N as its entry point at position 11 is the wire that had A as its entry point 13 positions earlier, since N is 13 places after A in the alphabet. Now $11 - 13 = -2$, and position -2 is the same as position $26 - 2$, i.e. 24.We therefore look up what letter A enciphers to at position 24. If this is, say, G then N at position 11 will encipher to the letter which is13 places after G in the alphabet at position 11, that is, to T.

Readers who are familiar with matrices will recognise that what we are doing, in effect, is representing the encipherment provided by a wheel as a 26×26 matrix. The first *column* gives the encipherment of the complete alphabet at position 1 and the first *row* gives the encipherment of A at each of the 26 positions.

The matrix can then be completely filled in from either its first row or its first column by using the 'diagonal property' explained above. A cryptographic feature of some importance is that whereas any *column* will contain all 26 letters of the alphabet, since two letters cannot encipher to the same letter at the same wheel position, the *rows* may contain one or more letters twice or more, since there is nothing to prevent a letter enciphering to the same letter at two or more positions. In fact, with a wheel of size 26, or any *even* number of contacts, it is certain that each row will contain at least one repeated letter. In the 6-letter example above this already occurs, C goes to A at positions 1 and 3. With an *odd* number of contact points it is possible that the rows will not contain any repeated letters. From a cryptographic point of view the fewer repeated letters in a row the better. (For an explanation of this, and further details see M14.)

Encipherment by the Enigma

We have just seen how a single wired wheel enciphers a letter. In the Enigma the current from the keyboard letter passes through the entry wheel and then through the three wheels R1, R2 and R3, after which it is 'turned around' by the reflector, U, and then goes back through the three wheels in the order R3, R2 and R1 before finally passing back through the entry wheel to light up the lamp which indicates the cipher letter. The original plaintext letter thus undergoes 9 changes before it finally emerges as a cipher letter; in fact, as we shall see later, in most military versions of the Enigma there were 2 further changes, making 11 in all.

If all the wheels were fixed the Enigma would merely provide a complex way of generating a simple substitution cipher, but they are not fixed. When a keyboard letter is pressed the rightmost wheel, R1, immediately turns one position and the current then passes through the machine. After 26 letters have been enciphered R1 will be back in its original position. Unless R2 or R3 had moved in the mean time the Enigma would be equivalent to 26 simple substitution ciphers; R2, however, will have moved. The notch ring on R1 moves with the wheel and so, some time during the 26 encipherments, the V-shaped notch will have reached the position immediately in front of a lever at the back of the machine opposite R1, this will allow the lever to engage with the V-shaped notch and this in turn will allow a lever opposite R2 to cause R2 to turn one position. Since R2 has now moved the encipherment alphabets will all be different from what they were 26 encipherments previously. R2 thus moves at least once in every 26 letter encipherments; in fact it moves slightly more frequently than that, for R2 also has a notch ring on it, and when *its* notch is opposite a lever behind R2 the third wheel, R3, is caused to move one position *and R2 itself is turned as well.* The consequence of all this is that the three wheels will not all have returned to their original positions until

$$26 \times 25 \times 26 = 16900$$

letters have been enciphered. Thus the Enigma machine provides an automatic way of using 16900 simple substitution ciphers in succession. So, for example, if the notch ring on R1 were so fixed that its notch caused R2 to turn when R1 was at setting Z as shown in its window, and likewise with the notch ring on R2, the successive positions of the three wheels when they are started at positions A, Y, Y (reading from left to right) will be

A Y Y,
A Y Z,
A Z A,
B A B.

This plethora of substitution alphabets provides good security but that is not the end of the story, for before 16 900 letters have been enciphered the three wired wheels can be removed and put back in a different order on the common axle. In the original Enigma there were only three wheels in the set which was provided with the machine and so they could be ordered in six ways. The number of available simple substitution alphabets was therefore

$$6 \times 16900 = 101400.$$

In fact, since R2 can be *started* in any of its 26 positions, including Z, even though it cannot move into position Z during normal operation unless R1 also has previously been at Z, there are $6 \times 26 \times 26 \times 26 = 105456$ possible starting positions and simple substitution alphabets.

Assuming that a cryptanalyst had such an Enigma he would therefore be faced with 105 456 possible wheel settings for the start of each message and this, in the days before computers, would appear to present him with an impossible task. If the cryptanalyst didn't have an Enigma available, and didn't know the internal wirings of the three wheels and reflector, the number of possibilities that he would have to try would be very much larger for there are

$$25!\ (\text{i.e. } 25 \times 24 \times 23 \times 22 \times \cdots \times 2 \times 1)$$

possible wirings of each wheel, and this number is greater than

$$10^{25}.$$

Three such wheels, therefore, can be wired in more than

$$10^{75}$$

ways. Furthermore, the cryptanalyst wouldn't know the internal wiring of the reflector and this multiplies the number of possibilities by a factor of more than

$$10^{12}$$

(for the calculation of this number see M15). Consequently, the cryptanalyst faced with messages enciphered on an Enigma with unknown wirings would apparently have to try more than

$$10^{87}$$

decryptions before being sure of success. Cryptographers, however, assume that the enemy will have acquired one of their machines on the first day of usage and so the assessment of the security of the original Enigma must be based upon the figure not of

10^{87}

but of

105 456

cases to be considered which, in 1923, before the invention of the computer, might have been considered adequate for a machine intended for purely commercial use. The German military, however, did not think so and insisted upon certain changes which improved the security considerably, the most significant of which was the introduction of

The Enigma plugboard

The military version of the Enigma included a plugboard at the front, below the typewriter keyboard. This plugboard had 26 sockets that could be connected in pairs by means of 13 short cables. The effect of this was to interchange pairs of letters at both the input and the output stages. So, for example, if A was connected to W by a cable then whenever the cipher operator typed an A it would go into the Enigma as W, and vice versa. Similarly a cipher letter which emerged from the final wheel, R1, as A would light up the W lamp and so be recorded as W. The number of ways of pairing the 26 letters of the plugboard is the same as the number of possible Reflector wheels, i.e.

more than 10^{12}

and since the pairings on the plugboard were changed frequently (daily at first and thrice daily from 1944) this increased the problem for the cryptanalysts, who were now faced with having to consider more than 10^{17} possibilities instead of 105 456.

The Achilles heel of the Enigma

The internal wirings of the three wheels and reflector on the military version of the Enigma were, at the understandable insistence of the military, different from those on the original civil version of 1923, so that possession of a 1923-version civil Enigma would not help the cryptanalysts. In addition the plugboard had been introduced. Even if a cryptanalyst had acquired a military-version Enigma the 10^{17} possibilities would seem to make decryption of even a single message impossible. If a million decrypts a second could be tried it would still take several thousand years to try them all; and in the 1930s there were no computers. How then did it come about that, in 1932, a method was found for decrypting Enigma messages?

The fundamental flaw which led to the decryption of Enigma messages was not due to the design of the machine itself, but to the method which was used by the Germans to send messages. As has been remarked before, if a cipher operator is sending a message on any cipher system with variable parameters he must somehow either let the recipient know the values of those parameters, or leave him to work them out. The latter situation rarely applies, although it may apply in the case of the Jefferson cylinder. In the case of the Enigma such an arrangement is quite impractical and the recipient must be provided with all the information needed for the decipherment of the message. What information does the recipient need? Assuming that both the Enigma sender and the recipient are 'on the same net', that is, they are using the same Enigma wheels, in the same order, and the same plugboard wirings, there is still the problem that the sender must somehow let the receiver know the positions of the three wheels at the start of the message, and this necessary information constitutes the 'indicator' in this case. The choice of these starting positions, or 'settings', should be 'randomly' chosen by the sender, for if a number of messages are enciphered with the same initial setting the cryptanalyst would be presented with a 'depth' and the messages might then be readable, even though the identity of the wheels and/or their initial settings might not be discovered. Since there are 17 576 possible starting positions for the three wheels the probability that any pair of messages will be given the same starting settings is less than 0.000 06 if the settings are chosen randomly but, in practice, cipher operators are liable to develop habits, such as choosing the three letters for the wheel settings from the same row of the keyboard, which make the probability of a depth much greater. Even with randomly chosen settings if a large number of messages are sent with the same machine set-up (same wheel order and plugboard) a depth may well occur by chance: 200 messages, for example, would make it more likely than not that at least one pair of messages will be in depth (for an explanation see M16). So, in order to reduce the possibility of depths occurring, the set-up needs to be changed sufficiently often that large numbers of messages on the same set-up are very unlikely to occur.

The method initially adopted by the Germans for letting the recipient know the starting positions of the three wheels was based upon the use of a common *ground setting* and the procedure for the sender was as follows.

(1) Choose a 'random' set of three letters for the actual wheel settings to be used for the encipherment of the message; let us suppose that this is FMZ (say).
(2) Look up the list provided to all users of the 'net' and note the ground setting for the day (or period, if the ground setting is changed more than once a day); let us suppose that this is BLE (say).
(3) Turn the wheels to the ground setting, BLE.
(4) Encipher the three letters of your chosen random setting *twice*; that is, encipher the six letters FMZFMZ and note the six resulting cipher letters; suppose that these are LOCWHQ (say).
(5) Turn the wheels to the chosen random setting, FMZ, and encipher the message.
(6) Precede the cipher text of the message with the six cipher letters of the enciphered double indicator, LOCWHQ.
(7) Transmit the cipher message with the enciphered double indicator at the front.

On receiving the cipher message the recipient would proceed as follows.

(1) Set the three wheels to the ground setting for the day or period, BLE.
(2) Type in the six letters, LOCWHQ, of the enciphered double indicator.

(Remember that it is a characteristic of the Enigma that encipherment and decipherment are equivalent. Thus, since F in the first encipherment became L so L would have become F, which is what happens when we decipher L.)

(3) This should yield the chosen random indicator *twice*, viz: FMZFMZ. If this does not happen there has been a mistake and the message may have to be re-sent; a dangerous situation which may cause a breach of security.
(4) If the six letters do decipher to the same repeated trigraph, FMZ in this case, turn the wheels to these positions and decipher the text of the message.

Note that the word '*twice*' is in italics. As we shall see, it was this operational procedure, introduced to ensure correct receipt of the three-letter settings, that proved to be the Achilles heel of the Enigma. We begin by noting that

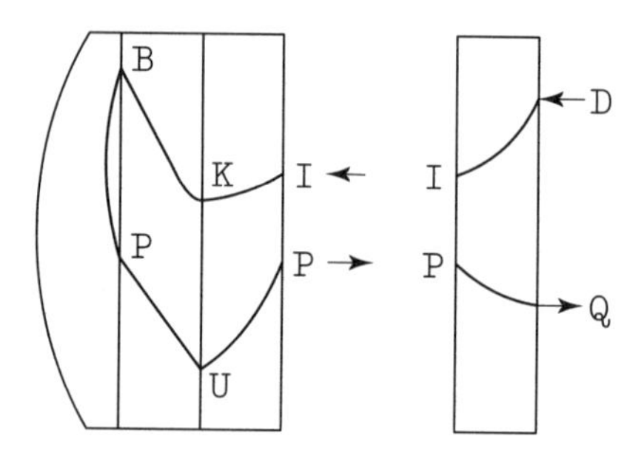

Figure 9. 3.

during the encipherment of the six-letter double indicator there is an 80% chance that only wheel R1 moves.

For wheel R2 can only move if R1 is on a notch position and on the original Enigma there was only one notch on the 26-position notch ring. Thus if the notch was effective when R1 was in position Z it could only cause R2 to move during the six-letter encipherment if the original setting of R1 was U, V, W, X or Y. (If the original setting of R1 was Z wheel R2 would turn immediately *before* the encipherment of the first letter and would remain there for the next five letter encipherments unless R2 itself was now at its notch position and so caused both R3 and itself to move at the next letter encipherment; this hardly affects the probabilities and, for simplicity, will be ignored. The cryptanalytic attack is easily modified to cover this unlikely possibility.) Therefore there are, normally, 21 of the starting positions of R1 which ensure that R2, and therefore R3, remain stationary throughout the encipherment of the six-letter indicator and so the probability that only R1 moves is 21/26, which is more than 0.8. Had there been more notches on the notch ring the probability would obviously have been lower. In the later stages of the War some notch rings had two notches and the notch rings of the Abwehr Enigma had many more, as we see later.

If R2 and R3 don't move during the encipherment of the six-letter indicator the Enigma becomes, effectively, a *one-wheel* machine, consisting of R1 and a composite reflector, made up of U, R3 and R2, none of which move during the encipherment. This is illustrated in Figure 9.3.

The composite reflector, formed by the stationary wheels R2 and R3 with the fixed reflector, U, is on the left. R1, the only wheel that moves during the six-letter encipherment, is on the right. The internal lines show how, in a typical encipherment, the plaintext letter D might become the cipher letter Q. The key point to note is that if, as shown in the diagram, D is enciphered to I and Q is enciphered to P by R1, then IP is a pair in the composite reflector, which means that during the six-letter encipherment any letter entering the composite reflector as I will emerge from it as P, and vice versa.

The indicator 'chains' in the Enigma

We have seen, in Figure 9.3, how the encipherment of a letter involves one of the 13 pairs in the composite reflector; in the typical example shown in Figure 9.3 the plaintext letter D was changed to I by R1 and so entered the composite reflector as I. After being changed to K then B then P then U respectively by R2, R3, U and R3 it emerges from R2, and hence from the composite reflector as P before finally passing back through R1 and producing cipher letter Q, since the entry wheel, being wired in alphabetical order, will leave the letter emerging from R1 unchanged. If, therefore, we knew the identity, internal wiring and starting position of R1 and *enciphered* D, to give I, and *enciphered* Q, giving P, we would discover one of the 13 pairs of the composite reflector. If we had a large number of such plain–cipher pairs we would find that if we had correctly identified both the wheel R1 and its initial setting we would always have got one of these 13 pairs of letters from the composite reflector. If, on the other hand, we had used the wrong wheel as R1, or the correct wheel but at an incorrect setting, we would have found not just 13 pairs but many more, since there are 325 possible pairings of 26 letters, and we would get contradictions such as 'IP is a pair but so is IM'. In this way we would soon know when we had found the correct wheel and its setting.

The problem is: *we don't know* the identity of the plaintext input letter (D in the example) although we *do* know the cipher output letter (Q); so how do we progress? The answer is that there is a feature of the six-letter indicators, originally discovered by the Polish cryptanalysts in 1932, that enables us to identify the correct R1 and its setting by this method, without knowing the input letters. The discovery relies upon two facts:

(i) plain–cipher pairs are reversible; e.g. if A enciphers to M (say) then M enciphers to A at the same wheel settings;

(ii) the six-letter indicator provides three cases where the same letter has been enciphered at positions in the text which are three apart so, in the example above where the enciphered indicator was LOCWHQ, (L, W), (O,

H) and (C, Q) are such pairs, and, *if R2 doesn't move during the indicator encipherment, we can hope to identify* R1 *and its setting.*

Since (i) is always true and (ii) holds about 80% of the time the prospects of success, using the method to be described, are good if enough information is available. Even if R2 *does* move the attack may still succeed for if R2 moves between the second and third letters (say) then the third and sixth letters (C and Q in the example) will have been produced by the same letter on a machine on which only R1 has moved. In the worst case, where R2 turns between the third and fourth letters, the attack is invalid, but the fact that it fails tells the cryptanalyst that R2 has probably moved and this may be helpful in a different attack.

Before describing the method for finding the identity and setting of R1 it is helpful to look at a small scale example. In this example the indicator consists of just one letter which is immediately repeated, rather than three letters repeated as in the Enigma itself. The cryptanalytic attack is the same in both cases. The chains are formed from the cipher pairs at positions 1 and 2 rather than those at positions 1 and 4 (and (2, 5) and (3, 6) for the Enigma).

Example 9.1 (Mini-Enigma)

The following 12 pairs of cipher letters are the result of enciphering the 12 plaintext letter-pairs AA, BB, ..., KK, LL (*in an unknown order*) at consecutive positions, and at a common ground setting, through a 12-letter Enigma-type cipher machine:

AK, BL, CI, DD, EB, FG, GE, HH, IA, JC, KJ, LF.

We form 'chains' from these pairs by joining pairs where the second letter of one pair is the same as the first letter of another pair, stopping when a letter is repeated:

AKJCI
BLFGE
DD
HH

We see that there are two chains of 5 different letters and two chains of just 1 letter. Is this a coincidence? No; it is not. The Polish cryptanalysts discovered that encipherment of pairs of the same letter at positions one or more places apart when only R1 moves will always produce chains which occur in pairs. The proof of this is not very difficult and the interested reader can find it in [9.1], but as further evidence here is part of a full-sized example with data generated on an actual Enigma machine. The example is worked out in detail in the article [9.1] referred to.

Example 9.2

From a batch of two-letter indicators, consisting of doublets enciphered on an Enigma at a common ground setting, the following set of 26, sorted by the first letter of the enciphered indicator, have been extracted:

```
AB BQ CD DK EZ FF GH HC IR JT KS LP ML
NU OO PI QN RA SJ TV UM VY WX XW YG ZE
```

Form the 'chains'.

Starting with `A` we have the chain `ABQNUMLPIR` of length 10.

Since `C` hasn't occurred we now start with that and find `CDKSJTVYGH`, also of length 10.

Since `E` is not in either of these chains we start with that and find the chain `EZ`, of length 2.

`F` is still missing and we see that it goes to itself, producing a chain of length 1, `F`.

`O`, `W` and `X` are the remaining letters and we see from the list that

> `O` goes to itself producing a second chain of length 1, `O`,
> and `W`, `X` go to each other, producing a second chain of length 2, `WX`.

In summary then we have:

Two chains of length 10: `ABQNUMLPIR` and `CDKSJTVYGH`.

Two chains of length 2 : `EZ` and `WX`.

Two chains of length 1 : `F` and `O`.

Had the 26 doublets not been enciphered at the same setting of the same wheel, R1, we would have had contradictions in the cipher doublets, having on thc onc hand `AB`, say, and also another pair, such as `AF`. The very existence of a unique set of 26 non-contradicting pairs supports the hypothesis that R1 was the same for all of them and nothing else moved.

In order to be able to obtain the chains we need sufficient messages to provide indicators beginning with each of the 26 letters of the alphabet. If the indicators are chosen at random many initial letters will occur twice or more, and so we would expect to need many more than 26 messages before we find 26 indicators beginning with the 26 different letters of the alphabet. How many messages might we need? It can be shown mathematically that we would probably need about 100 messages.

Ironically, if the cipher operators chose their indicators in a non-random manner, such as choosing all three letters from the same row of the keyboard, even more messages would be needed to get a full set of chains (M17). On the other hand the probability of finding two or more messages 'in depth' would be increased and the cryptanalysts might then be able to recover some plaintext, which might lead to the solution by a different route. Any kind of non-random feature in a cipher system or in its operational procedure might help the cryptanalysts.

Aligning the chains

The analysis that shows that chains of equal length occur in pairs also shows how to exploit this fact to discover the identity and setting of R1. To do this we must align pairs of chains of the same length, but *one of the chains of such a pair must be reversed.* Furthermore, we must try each possible alignment of the letters. Thus, for the two 10-letter chains above we can have 20 possible alignments, depending upon which chain we reverse and which letter comes first, viz:

```
ABQNUMLPIR
CHGYVTJSKD
```

is one possible alignment, but there are 19 others, for example

```
CDKSJTVYGH
IPLMUNQBAR
```

When we have the correct alignment of two related chains the *vertical* pairs of letters when encrypted through the correct R1 at its first setting will reveal pairs in the composite reflector whilst the NW–SE *diagonal* pairs when encrypted at the *second* setting of R1 will also reveal the same pairs. An incorrect wheel or incorrect settings will produce contradictions. This crucially important fact, a consequence of the operating procedure employed on the Enigma, was discovered by the Polish cryptanalysts in 1932; a proof is given in [9.1]. In this example we have some short chains and since there are fewer possibilities of alignment these would probably be tried first.

Identifying R1 and its setting

An example of the identification of R1 and its setting on a full size Enigma with known wheels but unknown plugboard would require a great deal of data and many pages of analysis, but that was the problem faced daily by the cryptanalysts. The method can, however, be illustrated using the data from the 12-point mini-Enigma example above. We assume that the plugboard is known and that we have to see if any of the known wheels at one of 12 possible starting positions could be R1. Since there are many incorrect possibilities only two cases will be examined: one incorrect and the other correct.

Example 9.3

The doublets in the 12-point mini-Enigma above are believed to have been enciphered at ***consecutive*** wheel positions on a common ground-setting with R1 having the encipherment table shown in Table 9.2.

Table 9.2

	Setting											
Input letter	1	2	3	4	5	6	7	8	9	10	11	12
A	K	A	G	L	H	F	I	C	F	D	L	E
B	F	L	B	H	A	I	G	J	D	G	E	A
C	B	G	A	C	I	B	J	H	K	E	H	F
D	G	C	H	B	D	J	C	K	I	L	F	I
E	J	H	D	I	C	E	K	D	L	J	A	G
F	H	K	I	E	J	D	F	L	E	A	K	B
G	C	I	L	J	F	K	E	G	A	F	B	L
H	A	D	J	A	K	G	L	F	H	B	G	C
I	D	B	E	K	B	L	H	A	G	I	C	H
J	I	E	C	F	L	C	A	I	B	H	J	D
K	E	J	F	D	G	A	D	B	J	C	I	K
L	L	F	K	G	E	H	B	E	C	K	D	J

Show that aligning the chains as

```
D    AICJK
H    BLFGE
```

(1) leads to contradictions if we assume that R1 is originally at Setting 1, but
(2) produces a solution if we take R1 to be originally at Setting 2.

Solution

(1) We begin by enciphering vertical pairs from the chain at Setting 1 and diagonal (NW–SE) pairs at Setting 2. The 12 pairs that we obtain should be consistent and provide the 6 pairs of the composite reflector: Table 9.3.

Table 9.3

Setting 1	Setting 2
DH→GA	DH→CD
AB→KF	AL→AF
IL→DL	IF→BK
CF→BH	CG→GI
JG→IC	JE→EH
KE→EJ	KB→JL

These produce several contradictions: eg Setting 1 implies that (A, G) are paired in the composite reflector whereas Setting 2 tells us that (A, F) are paired. Therefore R1 is not at Setting 1.

(2) We repeat the exercise but this time encipher the pairs at Settings 2 and 3: Table 9.4.

Table 9.4

Setting 2	Setting 3
DH→CD	DH→HJ
AB→AL	AL→GK
IL→BF	IF→EI
CF→GK	CG→AL
JG→EI	JE→CD
KE→JH	KB→FB

The two sets are in complete agreement and we conclude that we have identified R1 and that it was set at position 2 at the beginning of the encryption. We also now know that the pairings of the 12 letters in the composite reflector are

(A, L), (B, F), (C, D), (E, I), (G, K) and (H, J).

Since R1 and the actual reflector (U) are known the cryptanalyst would now try to find which combinations of U and two other wheels could produce these pairs. The original Enigma had only three wheels in its set and so, since R1 has been identified, there would be 'only' $2\times26\times26=1352$ possibilities to be examined. Whilst this may seem a large number it is small in comparison with the 105 456 cases which the cryptanalyst would have faced initially. Other lines of attack would now also be used, such as trying some possible plaintext beginnings for some of the messages.

It should be realised, of course, that the example gives only an indication of how messages on the original Enigma, with no plugboard and only three wheels in the set, could be decrypted. Many changes were made to the Enigma itself and to its operational procedures, particularly during the years 1938–45. For example:

(1) the repetition of the three-letter indicator was modified by the insertion of a pair of dummy letters and the use of a digraph substitution table;
(2) the use of a common ground setting was abandoned; operators chose their own ground setting which they gave, unenciphered, at the start of the cipher text;
(3) the three wheels were increased to five by all users and later to eight by the German Navy, who also used a four-wheel Enigma from 1942, which involved a new reflector.

These changes provided the cryptanalysts at Bletchley with a series of formidable challenges, which were overcome, in some cases quickly but for the four-wheel machine only with great effort.

For more information see [9.2] and [9.3].

Anyone interested in simulating the Enigma on a PC and following the steps of the original Polish solution would find [9.4] interesting.

For those who would like to try setting a (mini-) Enigma wheel here is:

Problem 9.1

On a 10-point mini-Enigma which has been used to encipher messages based on the digits 0–9 the results of enciphering the doublets 00, 11, ..., 99 at the same ground setting are obtained from the indicators (in unknown order) and are

$$(0,2), (1,6), (2,3), (3,9), (4,8), (5,5), (6,4), (7,7), (8,1) \text{ and } (9,0).$$

The first column of the encipherment table of the wheel which is believed to be R1 is

$$(0,8,6,4,3,7,1,5,9,2).$$

Complete the encipherment table and, with a suitable alignment of the chains, verify that R1 at setting 3 is consistent with the data. What are the pairings in the composite reflector?

Doubly enciphered Enigma messages

The question as to whether double encipherment is worthwhile was briefly discussed in Chapter 4 where, in effect, the answer was 'sometimes' since it is necessary to balance the increased security (if any) against the risk that the operator will use the ciphers in the wrong order, so producing a cipher message which cannot be read by the intended recipient but which may help the cryptanalyst. The Enigma provides a nice example which illustrates both these points. During the War some particularly important Enigma messages were doubly enciphered. An officer would encipher a message using the same wheels, but a different plugboard, as the cipher clerk; the wheel settings would be chosen from a special list of 26 possibilities, denoted by the letters of the alphabet. He would then give this cipher message to the clerk preceded by the word `OFFIZIER` and another word beginning with the letter indicating which of the 26 settings he had used. The clerk would then encipher this message in the normal way. This undoubtedly added to the security but on at least one occasion the two encipherments were applied in the wrong order (the officer and the cipher clerk were the same person on that day). Jack Good guessed what had happened and decrypted the message nevertheless. (See [2.4], Chapter 19, 159–60.)

The Abwehr Enigma

The Abwehr (Secret intelligence service of the German High Command) used a modified version of the Enigma which deserves special mention. There was no plugboard, which made life easier for the cryptanalysts, but the notch rings contained 11, 17 or 19 notches instead of 1 or 2. The rotors therefore moved much more frequently than in the 'standard' Enigma and the reflector also moved, which made life harder. The practice of enciphering the (four-letter) indicator twice at a given ground setting was retained and so an eight-letter group preceded the cipher text. The cryptanalysts still employed the 'chaining' method but, in addition, were able to exploit the situation where all four rotors moved simultaneously. For more details see [9.5].

10

The Hagelin cipher machine

Historical background

It might be thought, quite reasonably, that any country would try to keep secret the identity of its cipher machines and this is, in general, true. If the cipher machines were designed and built in the country concerned, as was the case with the Enigma in Germany, keeping their identity and details secret would be feasible, but if a machine was purchased from elsewhere it would be almost inevitable that others would eventually know about it. Under these circumstances it may seem surprising that before and during World War II there was a cipher machine that was used by several countries on both sides, including Germany, Italy, the UK, the USA and France. This machine was made in Sweden, a neutral country, by the firm of Boris Hagelin and was sold to anyone who wanted it. It was known by a variety of names in these various countries (Hagelin, M209, C36, C38, C41,...) but, with some variation, was essentially the same machine in all cases.

The basic function of the Hagelin machine was the provision of a long sequence of 'pseudo-random' numbers that were used as a key stream for the encipherment of plaintext by means of the equation

$$\text{(cipher letter)} = \text{key} - \text{(plaintext letter)} \pmod{26}. \qquad (10.1)$$

So, for example, if the plaintext letter was F (numerical equivalent = 5) and the key was 18 the cipher letter would be N since

$$18 - 5 = 13 \text{ (the numerical equivalent of N)}.$$

Note that equation (10.1) can be reversed, i.e.

$$\text{(plaintext letter)} = \text{key} - \text{(cipher letter)} \pmod{26}, \qquad (10.2)$$

so that there is reciprocity between plaintext and cipher letters, which means that encipherment and decipherment are identical, as in the Enigma. On the Hagelin however a letter can encipher to itself, which cannot happen on the Enigma.

Structure of the Hagelin machine

The Hagelin generated the key stream by means of six ***pinwheels*** and a ***lug cage***. The six pinwheels were mounted in parallel on a common axis and could be moved either independently, which was necessary when the machine was being set up, or together, which occurred every time a letter was enciphered or deciphered. Each wheel had a number of ***pins*** around its circumference and each of these could be moved so that it stuck out to either the left or the right side of the wheel. The number of pins was different for each wheel; looked at from the front of the machine the numbers of pins on the six wheels, reading from left to right, were 26, 25, 23, 21, 19 and 17. Every time a letter was enciphered each wheel moved one position. Since the six wheel lengths have no common factor they would not all be back in their starting position until

$$26 \times 25 \times 23 \times 21 \times 19 \times 17 = 101\,405\,850$$

letters had been enciphered. The letters of the alphabet were engraved around the rim of each wheel to enable the operators to set the wheels to their starting positions. On the 26-wheel the entire alphabet was engraved but the other wheels obviously required fewer letters. So, on the 17-wheel the letters A to Q sufficed.

The ***lug cage*** consisted of 27 horizontal bars arranged as a cylinder with the ends of each bar fixed into two circular discs. The cage could rotate about a common circular axis which was parallel to the common axis of the pin wheels. The cage was positioned immediately behind the pinwheels. On each bar of the cage there were 2 'lugs', small pieces of metal, which could be slid along the bar and fixed into any of eight positions. Six of the eight positions were directly opposite the six wheels; the other two, 'neutral', positions were not opposite any wheels but lay between wheels 1 and 2 and between wheels 5 and 6. If no more than 1 lug on *any* bar was opposite a wheel the machine was said to be 'in unoverlapped mode'; if *any* bar had 2 lugs opposite wheels the machine was said to be 'in overlapped mode'. The significance of this will become clear later.

In an unoverlapped cage one lug on each bar would probably be placed opposite a wheel. It was not essential for all 27 lugs to be so positioned, but there was no cryptographic advantage in using fewer. So, for example, the 27 lugs might be distributed among the six wheels as

26	25	23	21	19	17
4	1	9	6	5	2

The number of lugs opposite a wheel is sometimes referred to as the 'kick' of that wheel. Thus the 26-wheel above has a 'kick of 4' etc. The ordering is important; the cage above (4, 1, 9, 6, 5, 2) would not produce the same key stream as, say, (9, 1, 4, 2, 6, 5) although there would be some statistical similarity.

On the left hand side of the machine there was a small wheel engraved with the alphabet; this was used for the input of the plaintext letter. There was also a print wheel that printed the cipher letter on a thin strip of paper tape. This tape had gum on the reverse side, the purpose of which was to enable the operators to stick the cipher or plaintext onto sheets of paper. A reel of this gummed paper was housed at the back of the machine and there was also a notched screwdriver to enable the operators to set the pins and the lugs.

On the right-hand side of the machine there was a handle which, when rotated, would turn the cage and so encipher or decipher the text; this handle also caused each of the six wheels to move forward one position.

To encipher a letter the operator turned the input wheel until the chosen letter was opposite an arrow, and then turned the handle; the cage rotated, some bars on the cage shifted and caused both the input wheel and the print wheel to turn through anything from 0 to 27 positions. The cipher letter was printed and the six wheels all moved 1 position. Another letter could not be enciphered unless the input wheel had been moved by the operator. If by chance it were in the position where the next letter was opposite the arrow it was necessary to turn the wheel a few places forward and then the same number of places back. So, if the last cipher letter was, say, T and the next plaintext letter was also T the cipher operator would have to move the input wheel a few places and then turn it back to T before the handle would turn the cage.

For photographs of a Hagelin with its lid closed and open see Plates 10.1 and 10.2.

Encipherment on the Hagelin

Each wheel on the Hagelin had a number of pins sticking out to the left and the remainder to the right. The operators would be told, at the start

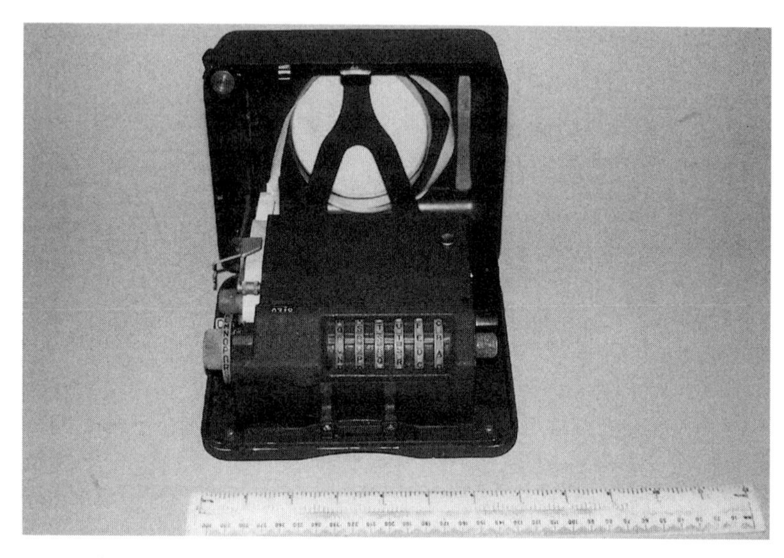

Plate 10.1 A Hagelin machine with top closed, ready for use. The positions of the six wheels are visible to the operator. The ruler in front shows the machine to be only about seven inches wide.

Plate 10.2 A Hagelin machine with its top open. The six wheels with some of the pins sticking out to the right and some to the left as well as some of the bars of the drum cage, with some lugs opposite wheels and some in neutral positions, can all be seen. The setting wheel and print wheel, letter counter and printer tape are also visible, on the left.

of each cipher period, to which side each pin was to be pushed. This was a tedious task since the total number of pins was

$$26 + 25 + 23 + 21 + 19 + 17 = 131.$$

A pin which stuck out to the left side of a wheel would have no effect on the key and so was an *inactive* pin; a pin sticking out on the right side would (in an unoverlapped machine) cause the key to increase by the number of lugs opposite that wheel on the cage, i.e. by the 'kick' of that wheel. Such a pin was called *active*. Ideally, there should be approximately equal numbers of active and inactive pins on each wheel; somewhere between 40% and 60% active would be about right.

At the start of each cipher period the operators would also have to move the lugs on each of the 27 bars so that the correct number were opposite each wheel, taking care that, in an unoverlapped machine, at most one lug on each bar was opposite a wheel, the other lug being left in one of the neutral positions.

When the input letter had been set opposite the arrow the handle was turned and the cage rotated. Each wheel would have a pin in position opposite some lugs; if the pin was on the left side of the wheel the lugs would pass it by but if the pin was on the right side the pin would cause the lugs to move which would shift the bars on which they were placed and each such move would cause the print wheel to move one place. Thus if there were 9 lugs opposite the 26 wheel, the print wheel would be caused either to move 9 places, if the pin opposite the cage was active, or not to move at all, if the pin was inactive. When the cage had completed its rotation the wheels would all move one position and so a different set of pins would now be in positions opposite the lugs.

Example 10.1

With the cage above, (4, 1, 9, 6, 5, 2), a typical keystream might be as shown in Table 10.1.

Table 10.1

26-Wheel	4	0	4	4	0	0	0	4	0	4	0	4	4	4	0	4	0	0	0	4
25-Wheel	0	1	1	0	0	0	1	1	0	1	1	1	1	0	0	0	1	0	1	0
23-Wheel	9	9	9	0	0	9	0	0	0	9	0	9	9	0	9	9	0	0	9	9
21-Wheel	6	0	6	6	0	0	0	0	6	6	6	0	6	6	0	0	0	6	6	0
19-Wheel	0	5	5	5	0	5	0	0	5	0	5	5	0	0	0	5	0	5	0	0
17-Wheel	2	0	2	0	2	2	0	2	2	0	0	0	2	2	0	0	2	2	0	2
Key	21	15	27	15	2	16	1	7	13	20	12	19	22	12	9	18	3	13	16	15

Note that the 19- and 17-wheels have made a complete cycle during the generation of these 20 key values so that their contributions to the total key begin to repeat, as indicated by the underlined values. Note also that since the key is interpreted by the print wheel (mod 26) the key value of 27 in the third position is effectively a key value of 1.

Example 10.2

Use the 20 key values in the example above to encipher the following text using the Hagelin method of encipherment

```
H A G E L I N X E N C I P H E R M E N T
```

We first convert the text into numbers

H	A	G	E	L	I	N	X	E	N	C	I	P	H	E	R	M	E	N	T
7	0	6	4	11	8	13	23	4	13	2	8	15	7	4	17	12	4	13	19

we then subtract these numbers from the corresponding key values (mod 26)

Key	21	15	27	15	2	16	1	7	13	20	12	19	22	12	9	18	3	13	16	15
Text	7	0	6	4	11	8	13	23	4	13	2	8	15	7	4	17	12	4	13	19
Cipher	14	15	21	11	17	8	14	10	9	7	10	11	7	5	5	1	17	9	3	22

Converting the cipher to letters and splitting into five-letter groups the cipher text would be transmitted as

```
OPVLR   IOKJH   KLHFF   BRJDW.
```

Choosing the cage for the Hagelin

In theory, the unoverlapped Hagelin cage may be any combination of six non-negative integers that add up to 27 or less. In practice, many of these cages would be very weak from a cryptographic point of view. Ideally, a cage should generate each of the possible key values, 0 to 25, equally often. This ideal is not achievable for, since each of the six wheels can be either active or inactive in any given position, there are

$$2^6 = 64$$

possible combinations of the six pins and hence 64 possible key values. Since all key values are interpreted (mod 26) the 26 possible values cannot all occur equally often because 64 is not a multiple of 26. On average, a key value can be expected to occur two or three times in a well-chosen cage. A cage such as (9, 9, 9, 0, 0, 0) is obviously very poor since the only key values that can be generated are 0, 1 (which occurs as 27), 9 and 18 and even these are non-uniformly represented, viz:

Key value	0	1	9	18
Number of occurrences out of 64 possible	8	8	24	24

Solving a message sent with such a cage wouldn't be very difficult since there are only four possible key values at any stage and two of these are much more likely than the other two. In the case of such a poor cage as this the process of solution is easily illustrated.

Example 10.3

The following is the cipher text of a message which has been enciphered on a Hagelin with cage (0, 0, 0, 9, 9, 9). X is used for spacing/punctuation. Decrypt the message.

```
ZCTAL   BRDSV   IBGDZ   SMFVM.
```

Solution

With such a cage the only possible key values are 0, 1, 9 and 18. If we subtract each letter of the cipher from these four values and write the resultant texts in four rows the decrypt must lie somewhere within the four rows. Since 9 and 18 are three times more likely to occur than 0 or 1 we would expect the majority of the plaintext letters to lie in the third and fourth rows. X is used for spacing/punctuation and, to make it more obvious, we replace it by ^ wherever it occurs.

We can save ourselves the tedium of subtracting the numerical equivalent of the cipher letters from the key values if we construct a table once and for all. This has the additional advantage that, unlike the book cipher tables of Chapter 7, the same table can be used for both encipherment and decipherment, because of the symmetry of the Hagelin encipherment/decipherment process mentioned above. When we have the table, given by Table 10.2, we simply look up the entries in the row corresponding to the cipher, or plaintext, letter and the column of the key value to obtain the plaintext, or cipher, letter, giving Table 10.3. The space marks are helpful and it is easy to pick out the plaintext, marked in **bold** in Table 10.3:

```
THIS IS A BAD CAGE
```

Table 10.2 *Encipher/decipher table for a Hagelin machine*

	Key value																									
	0	1	2	3	4	5	6	7	8	9	10	11	12	13	14	15	16	17	18	19	20	21	22	23	24	25
A	A	B	C	D	E	F	G	H	I	J	K	L	M	N	O	P	Q	R	S	T	U	V	W	X	Y	Z
B	Z	A	B	C	D	E	F	G	H	I	J	K	L	M	N	O	P	Q	R	S	T	U	V	W	X	Y
C	Y	Z	A	B	C	D	E	F	G	H	I	J	K	L	M	N	O	P	Q	R	S	T	U	V	W	X
D	X	Y	Z	A	B	C	D	E	F	G	H	I	J	K	L	M	N	O	P	Q	R	S	T	U	V	W
E	W	X	Y	Z	A	B	C	D	E	F	G	H	I	J	K	L	M	N	O	P	Q	R	S	T	U	V
F	V	W	X	Y	Z	A	B	C	D	E	F	G	H	I	J	K	L	M	N	O	P	Q	R	S	T	U
G	U	V	W	X	Y	Z	A	B	C	D	E	F	G	H	I	J	K	L	M	N	O	P	Q	R	S	T
H	T	U	V	W	X	Y	Z	A	B	C	D	E	F	G	H	I	J	K	L	M	N	O	P	Q	R	S
I	S	T	U	V	W	X	Y	Z	A	B	C	D	E	F	G	H	I	J	K	L	M	N	O	P	Q	R
J	R	S	T	U	V	W	X	Y	Z	A	B	C	D	E	F	G	H	I	J	K	L	M	N	O	P	Q
K	Q	R	S	T	U	V	W	X	Y	Z	A	B	C	D	E	F	G	H	I	J	K	L	M	N	O	P
L	P	Q	R	S	T	U	V	W	X	Y	Z	A	B	C	D	E	F	G	H	I	J	K	L	M	N	O
M	O	P	Q	R	S	T	U	V	W	X	Y	Z	A	B	C	D	E	F	G	H	I	J	K	L	M	N
N	N	O	P	Q	R	S	T	U	V	W	X	Y	Z	A	B	C	D	E	F	G	H	I	J	K	L	M
O	M	N	O	P	Q	R	S	T	U	V	W	X	Y	Z	A	B	C	D	E	F	G	H	I	J	K	L
P	L	M	N	O	P	Q	R	S	T	U	V	W	X	Y	Z	A	B	C	D	E	F	G	H	I	J	K
Q	K	L	M	N	O	P	Q	R	S	T	U	V	W	X	Y	Z	A	B	C	D	E	F	G	H	I	J
R	J	K	L	M	N	O	P	Q	R	S	T	U	V	W	X	Y	Z	A	B	C	D	E	F	G	H	I
S	I	J	K	L	M	N	O	P	Q	R	S	T	U	V	W	X	Y	Z	A	B	C	D	E	F	G	H
T	H	I	J	K	L	M	N	O	P	Q	R	S	T	U	V	W	X	Y	Z	A	B	C	D	E	F	G
U	G	H	I	J	K	L	M	N	O	P	Q	R	S	T	U	V	W	X	Y	Z	A	B	C	D	E	F
V	F	G	H	I	J	K	L	M	N	O	P	Q	R	S	T	U	V	W	X	Y	Z	A	B	C	D	E
W	E	F	G	H	I	J	K	L	M	N	O	P	Q	R	S	T	U	V	W	X	Y	Z	A	B	C	D
X	D	E	F	G	H	I	J	K	L	M	N	O	P	Q	R	S	T	U	V	W	X	Y	Z	A	B	C
Y	C	D	E	F	G	H	I	J	K	L	M	N	O	P	Q	R	S	T	U	V	W	X	Y	Z	A	B
Z	B	C	D	E	F	G	H	I	J	K	L	M	N	O	P	Q	R	S	T	U	V	W	X	Y	Z	A

Table 10.3

Text	Z C T A L	B R D S V	I B G D Z	S M F V M
Key = 0	B Y H A O	Z J ^ I F	S Z U ^ B	I O V F O
Key = 1	C Z **I** B P	A K Y J G	T **A** V Y **C**	J P W G P
Key = 9	K **H** Q J ^	**I S** G R O	**B** I **D** G K	R ^ **E** O ^
Key = 18	**T** Q Z **S** G	R B P **A** ^	K R M P T	**A G** N ^ G

As a slightly harder case try this.

Problem 10.1

The following message has been enciphered on a Hagelin machine with cage (0, 5, 5, 5, 5, 5)

CBZPC CJXWY CXSHN IQUSR.

Decrypt the message.

Since some cages are obviously very 'bad' (that is, weak from the cryptographer's point of view) this raises the questions:

(1) How many possible cages are there?
(2) How many of these are 'good' cages?

We can find the number of distinct possible cages exactly since the number is given by

> number of possible cages = the number of representations of 27 as the sum of six non-negative integers

and although there is no simple formula for calculating this number we can use a nice mathematical identity and a computer to discover that the number is 811. If we insist that each wheel must have at least 1 lug opposite to it, because a wheel with no lugs opposite it might as well not be there, the number reduces to 331. The same technique reveals that if we only use 26 or 25 lugs in the cage the number of possibilities, allowing 0 lugs opposite one or more wheels, reduces to 709 and 612 respectively, or to 282 and 235 when we disallow 0 lugs. We must remember however that since the six numbers of a cage may be permuted and so generate different key streams, although these streams will have similar statistical features, the number of possible cages that the cryptanalyst must consider is substantially larger. Thus, in the case of 27 lugs there are 201 376 possibilities, not 811; for further details see M18.

How many of these 331 distinct possible cages are 'good'? Clearly it depends upon what we mean by 'good'. If we take it to mean

> a 'good' cage is one that generates all possible keys in the range 0 to 25

then we can use a computer to find how many of the 331 do this. The answer is 113. Permuting the cage makes no difference to its quality: a 'good' cage remains 'good' and a 'bad' cage remains 'bad'. If we allow 0 lugs the 881 possible cages yield only 120 'good' cages; in other words, only 7 cages with a wheel with 0 lugs produce all 26 possible key values (mod 26). Of course even a 'good' cage may be ruined by highly asymmetric pin settings, such as 90% active, 10% inactive.

Problem 10.2

Determine which of the following 6 cages generate all key values, 0 to 25 (mod 26). For those cages which fail to do this find which key values are missing.

(a) 0, 2, 3, 4, 8, 10;
(b) 1, 2, 2, 3, 4, 15;
(c) 1, 2, 2, 4, 5, 13;
(d) 2, 3, 3, 3, 4, 12;
(e) 2, 3, 3, 3, 5, 11;
(f) 2, 3, 4, 4, 7, 7.

The theoretical 'work factor' for the Hagelin

When a cryptanalyst is faced with messages on a totally unknown Hagelin how many possibilities are there? This number is sometimes referred to as the *work factor* and represents the number of cases that would have to be tried if the cryptanalyst attempted a 'brute force' attack. For the type of Hagelin that we have been discussing so far (there are other optional features which complicate things further, as we shall see) there are two relevant factors:

(1) the number of possible cages;
(2) the number of possible pin settings.

Assuming that the cage uses 27 lugs, and that we allow wheels to have 0 lugs, there are over 200 000 ways of distributing them between the six wheels, so we'll take the first factor as 2×10^5. The second factor can be written down at once. There are 131 pins and each pin can be in either of two positions which means that there are 2^{131} possible pin settings. Since 2^{131} is more than 2.5×10^{39} the product of the two factors exceeds

$$5 \times 10^{44}.$$

Because the cryptanalyst can't be sure that there are 27 lugs being used he would have to be prepared to try other cages, such as those having only 26 or 25, which would approximately treble the work factor. It is clear from these numbers that a 'brute force' attack is out of the question. In fact cryptanalysts do not have to resort to 'brute force' attacks to solve the Hagelin, although solving the machine *from cipher messages alone* is difficult and requires a lot of text.

The situation is quite different if the cryptanalyst has managed to acquire a stretch of key and we shall see that given about 150 consecutive key values the solution is relatively easy. Once again then we will see that 'brute force' estimates can be very misleading when it is a question of assessing the difficulty of solving any cipher system. As we saw in Chapter

2, there are 26! possible simple substitution ciphers and the brute force work factor is about 10^{25} but such systems can be solved by hand in less than an hour, given 200 letters of cipher. Many types of cipher, including the Hagelin, are vulnerable to attacks involving mathematical or statistical weaknesses in the cipher itself or in its mode of operation, as in the Enigma. A cipher where there are no known weaknesses or 'trapdoors', which enable the cryptanalyst to break the system by a special route, would be formidable indeed. It is claimed that a modern system, the DES (Data Encryption Standard), is such a system. A brief description of this is given in Chapter 13.

Solving the Hagelin from a stretch of key

Recovery of the cage and pin patterns of a *non-overlapped* Hagelin is straightforward if we have more than 131 consecutive key values, as is shown below. 'How, though,' you might ask 'do we obtain 131 consecutive key values?' The answer is: 'By finding the plaintext of a message of length 131 characters or more.' This might be achieved in a number of ways including:

(1) finding that the same message has been sent in another cipher that is readable;
(2) finding that two or more Hagelin messages are in depth which may be readable by the method of crib-dragging as explained in Chapter 7;
(3) by the cipher operator making a mistake and having to retransmit the message on a slightly different set-up.
(4) by clandestine means, e.g. by an agent obtaining the plaintext of a message.

To illustrate the method of solution when a sufficiently long stretch of key has been obtained we look at the key that would be generated by a 'mini-Hagelin' with only three wheels. The corresponding attack for a full scale Hagelin undoubtedly involves more work but the method is the same and is based upon the fact that the key is the sum (mod 26) of the contributions of six wheels which move regularly and so, by subtracting the key stream from itself, suitably shifted, we can remove the contributions of any five of these wheels and so discover the 'kick' on the sixth wheel. Thus, if we generate a key by adding together the contributions of two mini-wheels, one of length 7 and with a kick of 5 and the other of length 9 and with a kick of 3 we might have

7-Wheel	5055005505500550550055 . . .
9-Wheel	0333003030333003030333 . . .
Sum	5388008535833553580388

We now make a copy of the key, shift it 7 places to the right (the length of one of the wheels) and subtract from the unshifted key viz:

	5 3 8 8 0 0 8 5 3	5 8 3 3	5 5 3 5 8	0 3 8 8
	5 3	8 8 0 0	8 5 3 5 8	3 3 5 5
Differenced key	0 0	−3 0 3 3	−3 0 0 0 0	−3 0 3 3

We note that

(1) all the differenced keys are multiples of 3, the kick on the 9-wheel,
(2) the differenced key pattern repeats after 9 positions, the length of the remaining wheel.

The process above is known as 'differencing at interval 7' and is usually symbolised in mathematics by using the Greek letter Δ ('capital delta'), the Greek equivalent of D, with a suffix denoting the appropriate interval, 7 in this case. So the process is fully symbolised by

$$\Delta_7$$

The process of differencing is also often referred to as 'delta-ing'.

There was no particular reason for first choosing to difference at interval 7, we could, of course, have differenced the key at interval 9, so let us do this:

Sum	5388008535	8	3	355	358	0	388
Shift 9 places	5	3	8	800	853	5	833
Subtract	0	5	−5	−555	−505	−5	−555

The pattern repeats at interval 7 and the kick on the 7-wheel is obviously 5.

We have clearly found the values of the kicks but what about the pin patterns on the wheels? These are found by examining the original key values once the kicks on the individual wheels are known. With a 'real' Hagelin there is the complication which we shall ignore for the present: when we see a key value of 0 or 1 it may really be a key of 26 or 27. In the mean time the method for recovering the pin patterns can be illustrated with another 'mini-Hagelin':

Example 10.4

The following stretch of key has been recovered from a mini-Hagelin with three wheels of lengths 5, 8 and 9. Recover the kicks and pin patterns.

6 4 3 6 8 6 4 5 1 8 4 1 8 6 8 1 6 3 9 0 4 6 8 6 0.

Solution

We may apply the differencing operations in any order. If we begin with interval 5 and then use interval 8 we will eliminate the contributions of the 5- and 8-wheels and should obtain a stream of numbers all of which are multiples (positive, negative or zero) of the kick on the 9-wheel. So:

Key	6	4	3	6	8	6	4	5	1	8	4	1	8	6	8	1	6	3	9	0	4	6	8	6	0
Shift 5						6	4	3	6	8	6	4	5	1	8	4	1	8	6	8	1	6	3	9	0
Δ_5						0	0	2	−5	0	−2	−3	3	5	0	−3	5	−5	3	−8	3	0	5	−3	0
Shift 8														0	0	2	−5	0	−2	−3	3	5	0	−3	5
$\Delta_8\Delta_5$														5	0	−5	10	−5	5	−5	0	−5	5	0	−5

The kick on the 9-wheel is obviously 5; note that the pattern begins to repeat after 9 places. Note also that the doubly differenced key has one value equal to *twice* the kick; this is a particular case of the following:

'When the key stream is differenced N times a value of up to $\pm 2^{(N-1)}$ times the kick may occur' (for an explanation see M19).

We now find the kick on the 8-wheel. This involves differencing at interval 5 and interval 9, in either order. Since we already have the key differenced at interval 5 (i.e. Δ_5 above) we need only difference that at interval 9:

Δ_5	0	0	2	−5	0	−2	−3	3	5	0	−3	5	−5	3	−8	3	0	5	−3	0
Shift 9										0	0	2	−5	0	−2	−3	3	5	0	−3
$\Delta_9\Delta_5$										0	−3	3	0	3	−6	6	−3	0	−3	3

The pattern begins to repeat after 8 places, as it should, and the kick on the 8-wheel is obviously 3.

Similarly, by differencing the original key stream at interval 8 and then at interval 9 we obtain

$\Delta_9\Delta_8$ 0 1 −2 1 0 0 1 −2

The pattern begins to repeat after 5 places and the kick on the 5-wheel is obviously 1.

We now have to find the patterns on the three wheels. We do this by looking at the key values and seeing how they might arise from combinations of the kicks (1, 3 and 5) on the wheels. This is easily done. With a full size Hagelin with six wheels it would not be so easy but the fact that the contribution of any wheel to the overall key repeats at intervals of the wheel length is a great help. In the case of this mini-Hagelin we can list the eight possibilities. Denoting an active pin by X and an inactive pin by O the only possibilities are shown in Table 10.4.

Table 10.4

5-Wheel Kick = 1	8-Wheel Kick = 3	9-Wheel Kick = 5	Key
O	O	O	0
O	O	X	5
O	X	O	3
O	X	X	8
X	O	O	1
X	O	X	6
X	X	O	4
X	X	X	9

Now we write down the key stream and the implied pin state for each wheel:

Key	6 4 3 6 8 6 4 5 1 8 4 1 8 6 8 1 6 3 9 0 4 6 8 6 0
5-Wheel	X X O X O X X O X O X X O X O X X O X O X X O X O
8-Wheel	O X X O X O X O O X X O X O X O O X X O X O X O O
9-Wheel	X O O X X X O X O X O O X X X O X O X O O X X X O

The patterns all repeat at their appropriate interval and so we can say:

Pattern on the 5-wheel is X X O X O
Pattern on the 8-wheel is O X X O X O X O
Pattern on the 9-wheel is X O O X X X O X O

With a full size Hagelin we would need a longer stretch of key of course. Every time we difference the key we lose a number of values, the number being equal to the length of the wheel at which we are differencing. Thus in the example above when we differenced at interval 5 the original 25 key values were reduced to 20 and when we then differenced at interval 8 we were left with only 12 values. For a full size Hagelin we need a minimum of 131 key values since 131 is the sum of the lengths of the six wheels. Rather more than 131 is desirable for we would like to have some additional key values in order to confirm that the final (five times differenced) values are repeating at the appropriate interval; something like 150 key values would be sufficient. In addition, the full size Hagelin has features which make the task of solving it significantly harder, as has already been indicated, and as we now see.

Additional features of the Hagelin machine

The Hagelin machine that has been described and analysed so far is the most basic type and we have ignored the fact that two of the possible key values, 0 and 1, are ambiguous and might really be 26 and 27 respectively. This means that if the stretch of recovered key contains any values which are 0 or 1, the cryptanalyst will have to consider 26 and 27 as alternatives when differencing. This could involve examining many alternative versions of the key stream; a failure of the differencing attack would indicate that one or more of the ambiguous values has been wrongly identified. In compensation the cryptanalyst does get some reward for correctly identifying a key of 0 or a key of 27 since the former implies that all six wheels are inactive at that point and the latter implies that they are all active; furthermore, a genuine key value of 1 implies that there is a wheel with a kick of 1, as does a key of 26 (we are assuming that all 27 bars are used, with no 'overlapping'). On balance, though, the task is made harder by these ambiguities.

On top of this complication, which applies to all models of the Hagelin, there are two additional features on the majority of models that add significantly to its security:

(1) the 'slide';
(2) 'overlapping'.

The slide

On Hagelin machines possessing this feature there is a small wheel on the outside which has the alphabet around its rim. This can be turned to any of the 26 positions and remains in that position whilst the message is being enciphered, or deciphered. The effect of the slide is to increase the key by a constant, so if the slide is E the key values are all increased by 4. In general the encipherment rule changes from

cipher letter = key − (plain letter) (mod 26)

to

cipher letter = (key + slide) − (plain letter) (mod 26),

the numerical value of the slide being the usual one, given in Table 1.1 viz: A = 0, B = 1, ... , Z = 25.

On machines that do not possess this feature the slide has a permanently fixed value: on the M209, for example, it is Z, which is numerically equivalent to 25, or −1 since all the arithmetic is (mod 26). So, on the M209, a key value which appeared to be 25 would in reality have been been generated by the six wheels as 0 or 26 and a key value which appeared to be 0 would have been generated by the wheels as 1 or 27. Identification of the slide would be a first step in key analysis.

Identifying the slide in a cipher message

The cryptanalyst may be able to identify the slide in a cipher message if he knows the cage that is being used. To do this he needs to compute a 'theoretical cipher distribution' and compare it statistically with the actual cipher letter frequencies in the message. For further details see M20.

The existence of the slide doesn't invalidate the differencing attack but it makes the initial recognition of the 0 and 1 key values more difficult. The slide would probably be changed for each message and the cipher operator would have to have some means of communicating its identity.

Overlapping

This feature was available on all models of the Hagelin. Recall that there are 27 bars and 54 lugs on the cage behind the wheels. On each bar the 2 lugs can be positioned opposite any of the six wheels or in one of the two 'neutral' positions. In an *unoverlapped* Hagelin one of the 2 lugs on each bar would be in one of the neutral positions. In an *overlapped* Hagelin 1 or more bars will have the 2 lugs opposite two of the wheels. This has the effect that where two wheels have a lug on the same bar their contribution to the overall key *when both are active* is 1, not 2. This is because the bars only move the print wheel 1 position irrespective of whether 1 or 2 lugs engage active pins. So, for example, if the 26-wheel has a kick of 5 and the 25-wheel has a kick of 6 and they share two bars where they both have an active pin then their combined active contribution to the total key is not 11 but 11 − 2 = 9. So

26- and 25-wheels both inactive – contribution to the key is 0;
26-wheel active, 25 inactive – " " " " is 5;
26-wheel inactive, 25 active – " " " " is 6;
26- and 25-wheels both active – " " " " is 9.

Overlapping will obviously affect the distribution of key values. Suppose we look again at the very poor cage (9, 9, 9, 0, 0, 0) but this time we overlap the kicks as follows:

(26, 25)-wheels overlap = 2;
(26, 23)-wheels overlap = 3;
(25, 23)-wheels overlap = 1.

Then the possible key values are given by Table 10.5, where X denotes an active pin and O an inactive pin, as before:

Table 10.5

26-Wheel	25-Wheel	23-Wheel	Key
O	O	O	0
O	O	X	9
O	X	O	9
O	X	X	17
X	O	O	9
X	O	X	15
X	X	O	16
X	X	X	21

Whilst this is still a very poor cage it does at least produce six different key values instead of four, which is all that the unoverlapped cage could produce; the key value of 9 is, however, still much the most common since it would occur 24 times out of 64 with the other five key values each occurring just 8 times. More importantly, though, the differencing attack would now fail since it would no longer be true that the contribution of a wheel to the total key would repeat at intervals of the wheel length. This is the main advantage of overlapping from the cryptographer's point of view.

Another advantage of overlapping is that the number of possible cages is enormously increased since there are now 27 *pairs* of lugs to be distributed instead of 27 *single* lugs. Does overlapping have any disadvantages? From a purely cryptographic standpoint probably not, providing that the overlaps have been assigned judiciously. The overlaps cause the key distribution to be changed and an unoverlapped cage that generates all possible 26 key values may cease to do so if it is overlapped. In addition, the cipher operators who have to set up the machine need to be particularly careful. Ensuring that the various wheels are overlapped to the correct extent with each other is vital; any mistakes may necessitate re-transmission of a message and the cryptanalyst will obtain useful information by comparing the two texts. The best way for the operators to be told the kick and overlap patterns of the wheels is to provide them with a 'map' showing the position of the 2 lugs on each of the 27 bars of the cage. The 'map' could either indicate the position of every lug on every bar, which would require eight columns – six for the wheels and two for the neutral positions; or it could show only the six wheel columns and, when a bar had only 1 lug shown, leave the operator to position the other lug in one of the neutral positions. The second method is used in the following

Example 10.5

The 'good' unoverlapped cage (9, 7, 5, 3, 2, 1) is modified to an overlapped cage (11, 9, 7, 5, 3, 1) where the first four wheels each have an overlap of 2 with the wheel to their right and the fifth wheel has an overlap of 1 with the sixth wheel. Draw up a suitable drum cage map for the cipher operators. Does the overlapped cage generate all 26 possible key values (mod 26)?

Solution

See Table 10.6.

Examination of the set of 64 key values generated by this cage shows that 5 key values 2, 4, 13, 20 and 25 cannot occur whereas the key value 17 occurs six times. It is not therefore a particularly satisfactory cage despite posing difficulties for a cryptanalyst because of the overlapping.

Problem 10.3

Find which pin combinations in the overlapped cage above produce the key value 17.

Solving the Hagelin from cipher texts only

Solving a Hagelin cipher message 'from scratch' requires a great deal of tedious work and I shall only give an indication of how the cryptanalyst would probably go about it. Detailed examples of the solution from cipher messages have been published and the interested reader should consult [10.4] or [10.5].

When Hagelin messages are first intercepted the cryptanalyst will have no knowledge of the cage or pin settings. Initially he may not even know that it *is* a Hagelin that is being used. If the cipher operators don't make mistakes the cryptanalyst is in for a lot of hard work. To have any chance of success he will need

Table 10.6

Bar	26-Wheel	25-Wheel	23-Wheel	21-Wheel	19-Wheel	17-Wheel
1	X					
2	X					
3	X					
4	X					
5	X					
6	X					
7	X					
8	X					
9	X					
10	X	X				
11	X	X				
12		X				
13		X				
14		X				
15		X				
16		X				
17		X	X			
18		X	X			
19			X			
20			X			
21			X			
22			X	X		
23			X	X		
24				X		
25				X	X	
26				X	X	
27					X	X

either one very long cipher message of several thousand letters,
or several messages of even greater total length.

He would then have to compute various statistics, beginning with an overall frequency count of the cipher letters, which would help to establish that a Hagelin machine was probably being used, for the non-uniform distribution of the 26 possible key values would be detectable in a sufficiently long text. If several texts had to be used it might be possible to determine their 'slides' relative to each other by calculating their cipher frequency correlation coefficients in pairs (if the machine was an M209, with fixed slide, this wouldn't arise).

The longest cipher text would now be written on a width of 17 columns and counts made of the cipher letters in each column. The object here is to attempt to put each column into one of two classes: those corresponding to an active pin and those corresponding to an inactive pin. If there is overlapping the inactive pin columns are not affected but the active columns are. By comparing the cipher letter frequency counts in the 'inactive' columns with the corresponding counts in the 'active' columns the cryptanalyst would hope to determine the kick on the 17-wheel. He might not succeed but whether he does or not he would carry out the same analysis on the widths of the other five wheels. If he were able to deduce one or more kicks with some certainty he would then see what the cipher distribution on the remaining wheels would be after taking the known kicks into account. Thus he would carry out an iterative process, hopefully steadily increasing the information about the cage and pin patterns. If there is overlapping the analysis is even more difficult, particularly if all of the wheels have some overlap. On the other hand, as the example above showed, too much overlapping may turn a 'good' cage into a poor one!

11

Beyond the Enigma

The SZ42: a pre-electronic machine

The Enigma and Hagelin machines provided a much greater degree of security than any earlier systems of encipherment other than the unbreakable one-time pad. The cryptographic principles on which these two machines were based were quite simple. The Enigma provided a large number of substitution alphabets whilst the Hagelin generated a very long stream of pseudo-random key. In theory either machine could be modified in order to make it even more secure. The number of wheels could be increased and in the Hagelin the wheels could be made longer. In practice, modification of an existing cipher machine may present major difficulties of manufacture, distribution and compatibility with the original machine, which may be vital. A four-wheel Enigma was, in fact, introduced in 1942 and compatibility with the original three-wheel version achieved by arranging that with the new components in specified positions the old and new versions were the same cryptographically. Several new models of the Hagelin were produced by that company in the 1950s with different sized wheels and other features, but these were genuinely different machines and no attempt was made to provide compatibility with the original.

It might seem obvious that increasing the number of components in, or increasing the complexity of, a cipher machine will make it more secure, but this is not necessarily so. The more components there are, the more likely it becomes that operators will make errors. The greater the complexity, the greater the chance of a machine malfunction. So it can happen that in attempting to *increase* the security of a machine the cryptographer might effectively *decrease* it. An example where increased

Plate 11.1 An SZ42 cipher machine on display at the Bletchley Park Museum, Milton Keynes. The setting rings for the 12 wheels and the wheels themselves can be seen at the front.

complexity gave rise to problems is given later. In addition, when cipher machines were mechanical devices, increasing the number of components would make the machine heavier and so less portable. The three-wheel Enigma weighed about 12 kilograms, a factor that had to be taken into account since the machines were intended for widespread use by operational units of all the armed forces. The four-wheel Enigma was used only by the German Navy and since it was carried only in ships and U-boats weight was not relevant.

If only a small number of cipher machines were required for some purpose, perhaps because they were to be based in permanent locations, then much larger machines could be considered. Such was the situation in 1941 when the German Army introduced a machine, subsequently known to them as SZ42, for communication between Vienna and Athens and, shortly afterwards, a somewhat similar machine, called T52, was used by the German Air Force.

A relatively small number of SZ42 machines were used, perhaps 52 on 26 links. The messages that they carried were of the highest importance and the machine was designed to provide an exceptionally high level of security. It was not intended to be a portable machine; it measured 20″ × 18″ × 18″ and was considerably bigger and heavier than the Enigma. See Plate 11.1 for a photograph of an SZ42.

Description of the SZ42 machine

The SZ42 contained 12 wheels of different sizes. Around the circumference of each wheel were pins, each of which could be placed in either of two positions that we can think of as 'active' and 'inactive'. Superficially this might seem to be a bigger version of a Hagelin machine but, as we shall see, it was quite different, both in the motion of its wheels and in its method of encipherment.

Table 11.1

Wheel	Length	Wheel	Length	Wheel	Length
A_1	41	B_1	61	C_1	43
A_2	31	B_2	37	C_2	47
A_3	29			C_3	51
A_4	26			C_4	53
A_5	23			C_5	59

The 12 wheels were arranged in three sets which will be referred to as A, B and C. Sets A and C each consisted of 5 wheels which we denote by (A_1, A_2, A_3, A_4, A_5) and (C_1, C_2, C_3, C_4, C_5) respectively. Set B had the remaining 2 wheels, to be referred to as (B_1, B_2). The wheel lengths were as shown in Table 11.1. The total number of pins is the sum of the wheel lengths, 501. The pins were changed daily; this must have been an onerous task for the cipher operators.

The wheels in set A and wheel B_1 each moved one place every time a letter was enciphered. Wheel B_2 moved one place if, and only if, the current pin on B_1 was active. The wheels in set C moved one place if, and only if, the current pin on B_2 was active. So although the wheels of set A moved regularly the wheels of set C moved irregularly.

Encipherment on the SZ42

The encipherment process in the SZ42 was quite different from that in the Hagelin. In the first place, the alphabet contained 32 characters, not 26. The 32 characters were those of the International Teleprinter Alphabet (ITA) which is ideally suited for use on five-hole tape where it can be used to represent the 26 letters and 6 functions, such as 'space' and 'carriage return'. Since $32 = 2^5$ the 32 characters can be written as combinations of five '0's or 1s', i.e. as five *bits*. The full ITA alphabet can be found

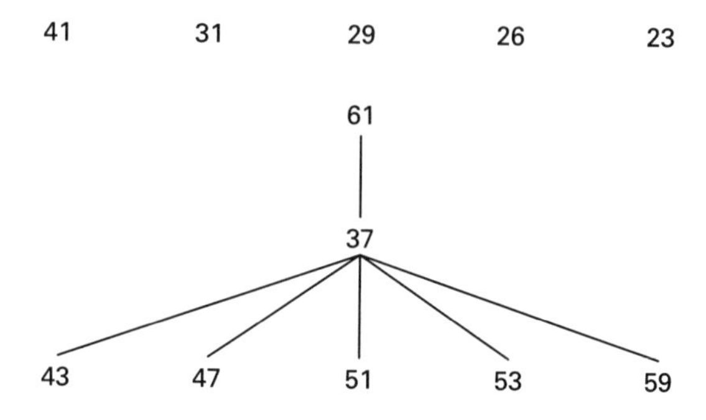

Figure 11.1. Motion control in the SZ42. The 5 wheels at the top and the 61-wheel moved every time a letter was enciphered. The 61-wheel controlled the 37-wheel which controlled all the 5 wheels at the bottom.

in many books such as [11.1], [11.2] but, as a sample, sufficient for our needs, we note that

A is `11000`
B is `10011`
...
P is `01101`
Q is `11101`
...
Z is `10001`

(there is no obvious relationship between representations of consecutive letters of the alphabet).

Although the 61- and 37-wheels determined the motion of the wheels in set C they played no direct part in the actual encipherment process, which involved only the five binary components of the plaintext letter and the current five pins of set A and the current five pins of set C.

A schematic diagram of the motion control in the SZ42 is shown in Figure 11.1.

The encipherment of a letter on the SZ42 was carried out as follows:

(1) the plaintext letter, P, was converted to its five-bit binary equivalent in the ITA code;
(2) the five bits of P were each enciphered separately;
(3) each bit of P was added (mod 2) to the value of the current pin on one of the wheels of set A, the value being 0 for an inactive pin and 1 for an active pin;

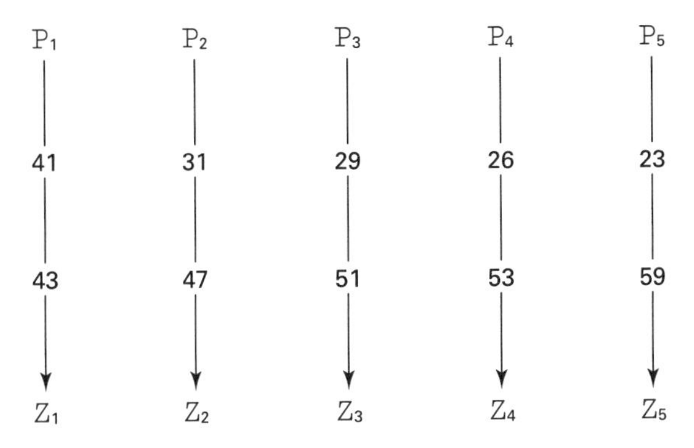

Figure 11.2. Encipherment process in the SZ42. The five streams of the plaintext letter, *P*, were separately added (mod 2) to the bits produced by the 2 corresponding wheels below to produce the five streams of the cipher letter *Z*.

(4) the five bits from stage (3) were added separately (mod 2) to the value of the current pin on one of the wheels of set C, the value being 0 or 1 as at stage (3);
(5) the resulting five-bit character, Z, was converted back, via the ITA code, to give the cipher letter on a printing mechanism;
(6) each wheel moved in accordance with the motion control mechanism.

The encipherment of each stream depended on just 2 of the 10 wheels: 1 from set A and 1 from set C. For example, the first stream was enciphered by the 41-wheel in set A and the 43-wheel in set C, whilst the fifth stream was enciphered by the 23-wheel in set A and the 59-wheel in set C.

A schematic diagram of the encipherment process on the SZ42 is shown in Figure 11.2.

As only (mod 2) addition of the plaintext and keys was involved the processes of decipherment and encipherment were identical since addition and subtraction are the same (mod 2).

As an illustration of the encipherment process:

Example 11.1
If the pin values on the wheels of sets A and C are

Set A 0 1 0 1 1
Set C 1 0 0 1 0

and the plaintext letter is `S` (=`10100` in ITA) what will be the cipher letter? Verify that decipherment yields the original plaintext letter.

Solution

Plaintext letter (S)	1 0 1 0 0
Set A pin values	0 1 0 1 1
(mod 2) sum	1 1 1 1 1
Set C pin values	1 0 0 1 0
(mod 2) sum	0 1 1 0 1 = P.

So the cipher letter will be P. If we now start with the cipher letter, P, then, with the same pin values the *decipherment* process gives

Cipher letter (P)	0 1 1 0 1
Set A pin values	0 1 0 1 1
(mod 2) sum	0 0 1 1 0
Set C pin values	1 0 0 1 0
(mod 2) sum	1 0 1 0 0 = S,

confirming the original plaintext letter.

Breaking and setting the SZ42

Assuming that the cryptanalyst knew the design details of the SZ42, as given in Figure 11.1, and that it was being used, how many possibilities would he have to consider before he could be certain of being able to decipher a message? The answer is easily obtained. The motion of the wheels and the encipherment are completely determined by the pins on the 12 wheels. Since there are 501 pins in all and each of them can be placed in either of two states, 'active' or 'inactive', the number of possibilities is

$$2^{501} \approx 10^{151}.$$

This is the '***breaking*** work factor' and it is so large that if every particle in the Universe was a computer and had been assigned full time ever since the 'Big Bang' to trying all the possibilities the solution would still not have been found. Clearly, a 'brute force attack' approach to breaking the SZ42 is hopeless.

If the cryptanalyst knew all of the pin patterns and a new message appeared he would have the 'easier' task of finding the settings of the 12 wheels at the start of the message. The number of possibilities that he would have to try is the product of the wheel lengths, viz:

$$23 \times 26 \times 29 \times 31 \times 37 \times 41 \times 43 \times 47 \times 51 \times 53 \times 59 \times 61$$

which is approximately 1.6×10^{19}. This is the '***setting*** work factor' and although it looks insignificant compared to the work factor for breaking it would still rule out a brute force attack even on today's most powerful computers, and in 1942 there were none.

The story of how SZ42 messages were nevertheless decrypted is described in some detail in [11.3]. The basic attack was to find the settings of the wheels of set A, using statistical analysis to find 'likely candidates'. Since even this involved more than 22 000 000 possibilities (the product of the wheel lengths of set A) a machine (known as Colossus) was designed and built for the purpose. Although intended for one very special purpose Colossus was in some respects an early electronic computer.

Problem 11.1

A cipher operator made a single error in setting one of the pins on one of the wheels in each of the following cases and the cipher messages were re-sent using the corrected pin settings. What would the cryptanalyst notice when comparing the two cipher messages in each case?

(1) The messages were enciphered on a Hagelin machine and the incorrect pin was on the 23-wheel;
(2) the messages were enciphered on an SZ42 and the incorrect pin was on the 31-wheel;
(3) the messages were enciphered on an SZ42 and the incorrect pin was on the 61-wheel.

With 131 pins on the Hagelin and 501 on the SZ42 such errors could easily occur.

Modifications to the SZ42

In order to increase the complexity of the SZ42 and so make life even more difficult for the cryptanalysts three modifications were introduced at different times. All of these were intended to make the stepping of the wheels of set B more unpredictable. Instead of the motion of the wheels of set B being controlled exclusively by the current pin value on the 37-wheel:

(1) the current pin value on the 31-wheel was added (mod 2) to the current pin value on the 37-wheel; or
(2) the current pin value on the 43-wheel was added (mod 2) to the current pin value on the 37-wheel; or
(3) the value of P_5 *two positions back* was added (mod 2) to the current pin value on the 37-wheel.

The third of these was soon abandoned since it required accurate transmission of the message, which could not always be guaranteed. If 'garbles' occurred the receiving operator might find that the decrypt became meaningless and so would ask for a re-transmission which might help the cryptanalysts. This illustrates the point made at the beginning of this chapter that increased complexity does not always guarantee greater security.

12

Public key cryptography

Historical background

The first general purpose computers were built in the 1940s. They were large, filling big rooms. They used hundreds of valves and consumed many kilowatts of electricity. They performed about a thousand instructions a second, which was considered amazing at that time, and they were popularly referred to as 'giant brains'. A few people, including Alan Turing, discussed 'whether machines could think' and laid bets as to whether a machine would defeat the World Chess Champion in the next 25 years. The former question remains a matter for debate; the latter was settled about 45 years later when a World Chess Champion *did* lose a match to a computer.

These early machines had very small direct access memories, only a thousand or so 'words', based upon cathode ray tubes or mercury delay lines. They rarely functioned for more than a few minutes before breaking down. Their input and output were primitive: paper tape or punched cards and a typewriter. They also cost a great deal of money; £100 000 in 1948 which was equivalent to several millions 30 years later. Very few people knew how to write programs for them. There was virtually no software (as it later became known) and all programs had to be written in 'absolute machine code'.

Even the instruction codes of these machines were very limited. The first machine at Manchester University in 1948, for example, had no division instruction [12.2], so division had to be programmed by repeated subtraction. Programmers, who were usually mathematics, science or engineering graduates, were very skilful and competed with each other at finding elegant and efficient ways of carrying out various processes, such as finding if the binary representation of a number contained just one non-zero bit. It was, of course, a very exciting era for those who were involved. It seemed that there was no need for large numbers of computers; one estimate was that six machines of the power of the Manchester computer would suffice for the whole world. It also seemed that programming computers would remain an esoteric art accessible only to an elite. It is doubtful if anyone could even imagine the tremendous advances that would transform the computing situation within a few years.

The changes came swiftly. Within a few years programmers, tiring of writing the same chunks of absolute machine code time after time, developed compilers for what became known as *higher level languages*. Some of these were specific to particular types of machines but before long others, such as FORTRAN, which stood for 'formula translation', became widely available. These made it possible for programs to be written by many more people, and in a much shorter time than hitherto, since large tracts of machine code could now be produced by writing a few lines in the higher level language. At the same time the technology was changing rapidly, cathode ray tubes and delay lines were replaced by core store memories which were faster and more reliable as well as providing far greater capacity in a lot less space. Valves were replaced by transistors and this soon led to integrated circuits with hundreds, then thousands, then millions of transistors on a few square centimetres of silicon which replaced bulky circuit boards. Within 20 years computer speeds had increased a thousandfold and memory capacities a hundredfold. There were myriads of programmers around the world, and countless numbers of others who used 'packages' for word-processing, spread-sheets or just 'games', on the computers without having any idea of the hundreds of thousands of lines of machine code that they were invoking.

Another major advance started in the 1960s when 'multi-access' computing, in which many people used teletypes connected to a mainframe computer, was developed and within a few years networks of computers were established, allowing people to have access to more than one machine and to communicate with others in distant places. The benefits were obvious and people in different continents not only sent 'electronic mail' to each other but also collaborated in research without ever actually meeting. The cost of computers also fell, not just in real terms but in absolute terms, by factors of 100 or more. All this rapidly became history and might have been lost in the mists of time but some accounts of the history of computing up to about 1980 were published and so preserved at least part of the story of what has been one of the most remarkable technological advances of all time; see, for example, [12.2], [12.3], [12.4] and [12.5].

Security issues

Security became an issue from the start. Initially it was mainly a question of *physical* security: protecting the equipment from physical damage such as fire or flooding. Within a few years however there were reports of disgruntled employees causing havoc by modifying or destroying key programs. One often cited case, possibly apocryphal but certainly credible, was of a programmer who had written a payroll program and who noticed that his employers had a habit of sacking their employees at very short notice. He therefore inserted a section of code into the program to check that his name was still on the payroll and, if it wasn't, to delete the entire program. In due course he was sacked and the payroll program deleted itself. The story went on to say that the programmer had to be re-hired at an enhanced salary to put things right. This may or may not be true but it draws attention to a serious possibility, known later as a 'Trojan horse' attack, and the need for a system of checking that programs had not been 'modified' which, in turn, led to the development of 'anti-virus software'. These problems still remain and computer viruses are regularly reported; many of these are just a nuisance but some are potentially disastrous.

Protection of programs and data

The integrity of programs and data can be protected either with or without encipherment. Programs consist of blocks of code. Data are usually stored, or transmitted, in blocks or *packets*. If every block is converted into a set of numbers which are then subjected to some mathematical function the final result can be either left as it is or enciphered. In either case the final value of the function can be stored, as a *check sum*, at the end of the block. If someone wishes to change anything in the block he must be able to compute both the mathematical function and, where relevant, its encipherment for the new block. If he can't do this the check sum will fail and it will be obvious that the block has been modified. An early, and very simple, version of this approach was the use of *parity checks* on programs and data stored on magnetic tape on computers in the 1950s. In this case there was no encipherment involved. A check sum, known as the *lateral parity check*, was automatically generated for every six-bit (later, eight-bit) character in the block and another check sum, the *longitudinal parity check*, based upon the number of 1s in each of the six (or eight) separate streams, was generated at the end of the block. These check sums were typically constructed so that each stream of 0s and 1s had an odd number of 1s in it. This was sufficient to detect and correct a single error in the block. If there were two or more errors they could not be corrected and, possibly, not even detected. The tape drives would carry out the checks automatically and re-read any block that failed the *parity test*. Dust on the tape could cause a misread and the tapes would frequently be seen to oscillate backwards and forwards several times before moving on. Since the computer itself generated the parity check bits there was no difficulty in changing the data if a genuine *single* error was detected. The parity checks were not designed to beat crooks, who might have managed to change both the data and the check sums; they were there only to protect the data from dust, including cigarette ash, and machine malfunctions. Smoking in the machine room, prevalent in the early days, was often found to be the cause of tape errors and was subsequently banned. Occasionally there was nothing wrong with the data on the magnetic tape and it was the checking circuitry in the tape drives themselves that was faulty. The computer operators soon discovered this and also found that, in the USA, a quarter-dollar inserted into the appropriate slot on the tape drive caused the checking circuits to be overridden.

When it became possible to connect computers together so that 'messages', which might be programs or data, could be passed between them the question of how to ensure that the messages did not get 'garbled', either accidentally or deliberately, en route had to be considered. Whilst parity checking, particularly if more sophisticated error correcting codes such as those described in [1.1], [1.2] and [1.3] were used, could provide protection against accidental corruption, a crook would be able to generate the appropriate check sums for any block that he had changed and so escape detection. It thus came to be realised that encipherment, using a system which a crook couldn't replicate, was necessary.

Encipherment of programs, data and messages

By the beginning of the 1970s most major businesses, government offices and academic institutions were using computers, and computer networks were spreading. A typical installation would have one or more large machines, *mainframes* as they were called, with numerous teletypes or, later, graphics terminals communicating with them by telephone lines. Many of these organisations were also using computers at distant sites as well as those 'in house'. In order that diverse users could communicate securely with each other some common form of encryption was required. Since all the users would have to know how the encryption would be carried out it was clear that the encryption algorithm would have to be made public and that the individual users would have to have their own, secret, keys, without which it would be impossible to decipher their messages. This, in turn, implied that the method of encryption must be extremely secure.

In addition to the problems already mentioned with the introduction of computer networks new aspects of security arose. Here are two examples:

(1) 'The authentication problem'. A user, X say, receives an e-mail message which apparently comes from Y. How does X know that the message really has been sent by Y and, even if it has, that it has not been altered in some way? The fact that the message has Y's e-mail address on it is no guarantee, since someone might be using Y's computer in his absence. Even if Y has a password it is possible that, having logged on, he has gone out of the room for a few minutes leaving his computer idling, a bad habit which invites misuse. This would enable a third party, Z, to use Y's computer in his absence to send the message to X. If X is Y's bank and the message is authorising the transfer of a large sum of money from Y's account to an overseas account the bank needs to have some way of checking that the message is genuine, otherwise a fraud can be successfully committed. Alternatively, is it possible that Y has indeed sent a message to X and that Z has somehow intercepted it and changed part of it so that it benefits him?

(2) 'The signature verification problem'. A user X sends a message to Y authorising some action or other. Subsequently X denies that he sent the message to Y. The dispute goes to a third party, a judge perhaps, who has to decide if X really did send the message or not. Is there a way in which X, having signed the message, cannot subsequently deny that he sent it and, conversely, that Y cannot claim to have received a different message?

These are practical problems which have been the subject of much research and discussion ever since computer networks came into existence and we shall return to them in the next chapter. Various solutions have been proposed but the essential feature of most of them is that there is a method whereby two people who wish to communicate are enabled to do so by means of a common encryption system which involves the use of one or more keys which only they know. The problem is 'How are they to let each other know their secret key(s) without other people discovering them too?'

The key distribution problem

The situation is that X and Y wish to communicate with each other using an agreed encryption system. A third party, Z, knows their agreed encryption system, is able to intercept their messages and would like to be able to read them. X and Y may or may not know of the existence of Z but they want to be sure that their messages should be unintelligible to anyone other than themselves. The system, which they must assume is known to Z or to anyone else, requires the use of one or more keys, which they intend to keep secret but which may be changed from time to time, possibly with each message, or possibly less frequently. Anyone who gets hold of the keys and knows the method of encryption can decipher their messages. It is therefore essential that the keys remain secret, but how can X and Y tell each other their keys without running the risk that Z will intercept and be able to exploit them?

The Diffie–Hellman key exchange system

An elegant solution to the key exchange problem was proposed by Diffie and Hellman in 1976 [12.6]. Their method is implemented by X and Y as follows.

(1) X and Y agree upon the use of two integers p and m (say) where p is a large prime and m lies between 1 and $(p-1)$. The values of p and m need not be kept secret.

(2) X chooses a secret number, x, and Y chooses a secret number, y. Both x and y lie between 1 and $(p-1)$ and neither should have any factor in common with $(p-1)$. In particular, since $(p-1)$ is even, neither x nor y should be even. X and Y do not reveal their secret numbers to each other or to anyone else.

(3) X computes the number

$$k_x = m^x \pmod{p}$$

and sends it to Y who raises it to the power y, giving the number $(k_x)^y$.

Y computes the number

$$k_y = m^y \pmod{p}$$

and sends it to X who raises it to the power x, giving the number $(k_y)^x$.

(4) Now $(k_x)^y = (k_y)^x \equiv m^{xy} \pmod{p} = K$ (say) and K is the common key which both X and Y can use, even though neither of them knows the other's secret key.

Before one can use the Diffie–Hellman system it is necessary to be able to find a very large prime number, and this is a non-trivial task. We meet this problem again in the RSA encryption system (where two large primes are needed), where references to an interesting approach will be found.

In a realistic example the prime p would be very large but the essence of the method can be illustrated with a prime of moderate size.

Example 12.1

In the Diffie–Hellman system if $p=59$, $m=3$, $x=7$ and $y=11$ what will be the values of k_x, k_y and K?

Solution

(1) We first note that $(p-1)=58=2\times 29$ and so has no factor in common with x or y.

(2) X computes $3^7 \pmod{59}$ and Y computes $3^{11} \pmod{59}$. These calculations can be done in various ways, some more efficient than others (M22). In this case the numbers are sufficiently small that the powers can be handled on a pocket calculator. Thus

$$3^7 = 2187 = 37\times 59 + 4,$$
$$3^{11} = 177\,147 = 59\times 3002 + 29$$

so $k_x = 4$ and $k_y = 29$.

(3) The value of the common key, K, is then given either by $4^{11} \pmod{59}$ or by $29^7 \pmod{59}$. These should give the same value; if they don't we have made a mistake. Therefore as a check we compute them both. The numbers this time are somewhat bigger so we compute them by forming powers and reducing (mod 59) at each stage thus:

(i) $4^5 = 1024 = 17\times 59 + 21 \equiv 21 \pmod{59}$

so

$$4^{10} \equiv 441 = 7\times 59 + 28 \equiv 28 \pmod{59}$$

and therefore

$$4^{11} \equiv 4\times 28 = 112 = 1\times 59 + 53 \equiv 53 \pmod{59}.$$

We conclude that the common key, K, is 53.

(ii) We check this by computing $29^7 \pmod{59}$.

$29^2 = 841 = 14\times 59 + 15 \equiv 15 \pmod{59}$

so

$$29^3 \equiv 29 \times 15 = 435 = 7 \times 59 + 22 \equiv 22 \pmod{59}.$$

Squaring,

$$29^6 \equiv 484 = 8 \times 59 + 12 \equiv 12 \pmod{59}.$$

Finally,

$$29^7 \equiv 29 \times 12 = 348 = 5 \times 59 + 53 \equiv 53 \pmod{59}$$

and we have confirmed that k = 53 in this case.

The reason for the restriction that neither x nor y has a factor in common with $(p-1)$ is that if x, say, had such a factor the value of k_x, and hence of the common key K, could turn out to be 1, irrespective of the value of y, which is cryptographically undesirable. For example if $p = 31$ and $m = 2$ neither x nor y should have a factor in common with 30. If X were to choose $x = 5$, say, then

$$k_x = 2^5 = 32 \equiv 1 \pmod{31}$$

hence $k_x = 1$ and so $K = 1$ no matter what value Y chooses for y.

Strength of the Diffie–Hellman system

How secure is the Diffie–Hellman system? It must be assumed that an interested third party, Z, can obtain the values of m, p, k_x and k_y but not of x or y. The security therefore depends upon how difficult it would be for him to obtain the value of x, say, given the value of k_x ($= m^x \pmod{p}$). This is known to be an extremely difficult problem (the discrete logarithm problem) unless the prime, p, is of a special form. In general it can be considered impossible if p is larger than about 10^{200}. A related problem arises with the RSA encryption method as we shall see.

There is, however, an alternative attack which Z could employ if he can intercept messages en route between X and Y and slightly delay them. Since he knows the values of m and p he might be able to retain the values of k_x and k_y and substitute a value of his own, say $k_z = m^z$, and send it to both X and Y. X and Y would then unwittingly use k_z in their encryptions and Z would read their messages. He would then re-encrypt the messages using the original key k_x or k_y as appropriate and X and Y would be unaware that their messages were being read.

There is considerable interest in ways of preventing attacks such as this see; for example, [12.7].

Despite this potential weakness the Diffie–Hellman system provides a method for key exchange which would defeat all but the most determined and well-equipped adversary. In particular it can be used as the starting point for using encryption systems such as DES, which we deal with later.

13

Encipherment and the internet

Generalisation of simple substitution

In a simple substitution cipher the letters of the alphabet are replaced by a permutation of themselves. We have seen that such a cipher is easily solved, given as few as 200 letters, by counting their frequencies and using knowledge of the language. To use such a cipher simply requires a 26-long table of the permuted alphabet. If, say, A was replaced by R, N by C and T by H then AN would become RC in the cipher and AT would become RH. Thus R, the substitution for A, would appear both times.

Since a simple substitution cipher replaces *single* letters by the same letter each time, irrespective of whatever letter precedes or follows them, the frequency count attack will ultimately succeed. To counteract this if we had a system where the encipherment of a letter depended on some of the letters on either side of it then AN might encipher to RC whereas AT might encipher to, say, KW and the monograph frequency count method would fail. Such a system *could* be based upon a substitution table which listed all 676 $(=26\times26)$ digraphs and their cipher equivalents. Effectively we would have a two-part code-book; the first part would list all 676 *plaintext* digraphs in alphabetical order on the left of the pages with their cipher equivalents listed opposite them on the right. The second part would list the 676 *cipher* digraphs in alphabetical order on the left with their plaintext equivalents on the right. This system, which one might call *digraph substitution*, would be more secure than simple substitution but would be more tedious to use and the user would need to have the two tables readily available, unless he had an exceptional memory.

If the cryptographer were prepared to have two tables on hand, listing the 17 576 ($=26\times 26\times 26$) plaintext and cipher trigraphs, an even more secure system, *trigraph substitution*, could be used.

Clearly, one could devise increasingly secure systems in this way but, in practice, the tables would be unwieldy and even a system based on *strings* of 4 letters (i.e. tetragraphs) is hardly practicable.

Suppose, however, that a system could be devised which took a string of letters of fixed length and somehow converted them automatically into another string and which guaranteed that changing *any* of the letters in the first string would produce an entirely different second string. Such a system would not require printed tables and could be very secure indeed, depending upon the method by which the string is converted and the number of letters in the strings. At the lower end of the scale of such systems we have the Julius Caesar method: the conversion being done by moving each letter three places forward in the alphabet and the fixed length of the string being 1. At the top end of the scale we have a method known as RSA, after Rivest, Shamir and Adelman who devised the method in 1978 [13.1], which can be used to encipher very long strings, e.g. 100 letters, and which provides an extremely high degree of security. This may seem remarkable enough but even more remarkable is the fact that the RSA is *a public key system*, which means, as was explained in Chapter 12, that the details of how to *encrypt* a message are made publicly available, but only the 'owner' of the public key knows how to *decrypt* the messages which are sent to him. The owner can, however, reply to his correspondents by encrypting *his* messages in such a way that they *can* decrypt them.

Although in theory RSA encryption could be carried out by hand the computations, which involve modular arithmetic with very large integers, could only realistically be carried out on a computer with facilities for multi-length arithmetic.

Factorisation of large integers

It is a relatively easy task to multiply two numbers together, particularly if we have a calculator available, provided that they are not too large. If the numbers are both less than 10 a child should be able to do it unaided. If they are both less than 100 most people, hopefully, could get the answer with paper and pencil. If both numbers are bigger than 10 000 it is likely that a calculator would be employed.

The opposite problem, finding two or more numbers which when multiplied together produce a given number, is called *factorisation*. This is a much harder task than multiplying numbers together, as anyone who has tried it knows. For example: if we are asked to multiply 89 by 103 we should be able to get the answer, 9167, in less than a minute. If, however, we were asked to find two numbers whose product is 9167 it would probably take us considerably longer. How would we do it?

The standard method of factorisation

If we are asked to factorise a large number, N, we should use the fact that unless N is a prime number it must have at least two factors and the smaller of these cannot exceed the square root of N. This means that in the case where $N=9167$ we need only test for divisibility by the primes less than $\sqrt{9167}$, which is nearly 96. The largest prime below 96 is 89, so in this case we would succeed on the very last test, by which time we would have carried out more than 20 divisions. Had we carried out the tests on $N=9161$ we would have failed to find a divisor, since 9161 is a prime.

As N increases so does the number of tests that we have to make. Thus if $N=988\,027$ N either is a prime or is divisible by some number less than $\sqrt{988\,027}$, which is a little less than 994. We would then try dividing 988 027 by each prime less than 994. If we find a prime that divides 988 027 exactly, i.e. without leaving a remainder, we have solved the problem. If no such number is found we would know that 988 027 is prime. In fact

$$988\,027=991\times 997$$

and since both 991 and 997 are prime numbers the factorisation is complete. It would have taken us quite a lot of effort to do this because there are more than 160 primes less than 991, and we would have had to try them all before we were successful. Even with a calculator this would be a time-consuming and tedious job. Someone who has a computer and can program could, of course, get the computer to do the work. Irrespective of how it was done, by increasing N from 9167 to 988 027, a factor of about 108, we, or the computer, were faced with an increase in the number of divisions from about 20 to over 160. Note that although N increased by a factor of over 100, so that $\sqrt{N}$ increased by a factor of more than 10, the number of tests increased by a factor of only about 8. The explanation for this can be found in M21.

This method of finding the prime factors of a number by dividing the number by each prime less than its square root is, essentially, due to Eratosthenes and is the standard method both for factorising (if it works) and for showing that a number is a prime (if it fails). This is not the only method that might be used, sometimes a short-cut can be found; for example, someone might notice that

$$9167 = 9216 - 49 = (96)^2 - (7)^2 = (96 - 7)(96 + 7) = 89 \times 103$$

and, even better, that

$$988\,027 = 988\,036 - 9 = (994)^2 - (3)^2 = (994 - 3)(994 + 3) = 991 \times 997$$

but, in general, we are not so lucky. Sometimes, such as when the number that we are trying to factorise has a particular form such as

$$2^p - 1$$

where p is a prime number, there are special techniques which reduce the number of possibilities, but in the type of number which is relevant to the RSA system these special techniques are not applicable.

The RSA system of encryption, which is described below, relies for its *security* upon this fact: that it is very time-consuming to factorise a large number even if we are told that it is the product of two large primes. As for the *encryption process* in the RSA system the basis of this is an elegant and powerful theorem stated, without proof, by the French mathematician Pierre Fermat early in the seventeenth century. This is often referred to as 'Fermat's Little Theorem' and is not to be confused with the notorious 'Fermat's Last Theorem', which he also stated without proof, and which was not proved until 1993 [13.2]. Fermat *may* have had a proof of his 'Little Theorem'; it is extremely unlikely that he had a proof of his 'Last Theorem'. The Swiss mathematician Leonhard Euler gave a proof of Fermat's Little Theorem in 1760 and also generalised it, so giving us what is known as the Fermat–Euler Theorem and it is this that is used in the RSA encryption/decryption process.

As a preliminary it is instructive to look at some examples of

Fermat's 'Little Theorem'

Here are some elementary exercises.

What is the remainder when

(1) 2^4 is divided by 5?
(2) 3^4 is divided by 5?
(3) 3^6 is divided by 7?
(4) 5^6 is divided by 7?
(5) 3^{10} is divided by 11?
(6) 8^{10} is divided by 11?
(7) 59^{96} is divided by 97?

Solutions

(1) $2^4 = 16 = 3 \times 5 + 1$. Remainder is 1.
(2) $3^4 = 81 = 16 \times 5 + 1$. Remainder is 1.
(3) $3^6 = 729 = 104 \times 7 + 1$. Remainder is 1.
(4) $5^6 = 15625 = 2232 \times 7 + 1$. Remainder is 1.
(5) $3^{10} = 59049 = 5368 \times 11 + 1$. Remainder is 1.

(6) In this case we avoid having to deal with large numbers by using modular arithmetic, as described in Chapter 1:

$$8 = 2^3$$

so

$$8^{10} = 2^{30} = (2^5)^6 = (32)^6$$

and

$$32 \equiv -1 \pmod{11}$$

so

$$(32)^6 \equiv (-1)^6 = 1 \pmod{11}$$

so the remainder is 1.

(7) The remainder is again 1. Since $(59)^{96}$ is a very large number, containing 171 digits, it is even more essential to use modular arithmetic. For the details and method of calculation see M22.

Is the remainder always 1 when the exponent on the left is 1 less than the modulus on the right? No, it is not. Consider

$$2^{14} = 16\,384 = 1092 \times 15 + 4.$$

In this case the remainder is 4, not 1. Why is this? The answer is that $15 = 3 \times 5$ and so is not a prime and Fermat's Little Theorem only applies when the modulus is a prime, such as 5, 7, 11 or 97, as in the examples above. Euler showed how the theorem must be modified when the modulus is not a prime. The original theorem though is

Fermat's Little Theorem

If p is a prime number and m is any number which is not divisible by p then

$$m^{(p-1)} \equiv 1 \pmod{p},$$

i.e. $m^{(p-1)}$ leaves remainder 1 when divided by p.

The proof is not difficult and generalises fairly easily to give a proof of the Fermat–Euler theorem, which is given in M23.

The generalisation discovered and proved by Euler applies to *any* modulus but the version required by the RSA system requires only the case where the modulus *is the product of just two distinct primes* and so is worth stating in its own right:

The Fermat–Euler Therorem (as needed in the RSA system)

If p and q are different prime numbers and m is any number which is not divisible by p or q then

$$m^{(p-1)(q-1)} \equiv 1 \pmod{pq}.$$

In the example above we had $p = 3, q = 5$ and $m = 2$ and the theorem tells us that

$$2^{(2)(4)} \equiv 1 \pmod{15}$$

and indeed $2^8 = 256 = 17 \times 15 + 1$.

Encipherment and decipherment keys in the RSA system

To *encipher* a text by means of the RSA method we require:

(1) a large number n $(= pq)$ which is the product of just two distinct primes, p and q (The question as to *how* one finds very large primes is highly relevant. We have met this problem before, in connection with the Diffie–Hellman key exchange system. In general a considerable amount of computation is required. Since the primes to be used

should, in this case, not be of any special form, such as $2^p - 1$, there are no really fast methods; there is, however, a method for finding numbers which are *probably* primes, where the probability can be brought arbitrarily close to 1, i.e. to *near certainty*; for details see M24);

(2) an integer, e, known as the *encipherment key*, which has no factor in common with $(p-1)(q-1)$ or with n itself;

To *decipher* a text which has been enciphered using the RSA method we further require

(3) an integer, d, known as the *decipherment key*, which satisfies the condition that

$$ed \equiv 1 \pmod{(p-1)(q-1)}.$$

At this point it is worth noting that the relationship between e and d is symmetric which tells us that if d is the decipherment key for e then e is the decipherment key for d. The significance of this will become clear when we see how the 'owner' of the decipherment key, d, can reply to his correspondents.

When n and e are chosen d has to be found. There is a method for doing this *if p and q are known*; but if p and q are not known d cannot be found, and this fact is the basis of the security of the RSA system. If p and q are extremely large, bigger than 10^{200} for example, finding their values in a reasonable time, which means factorising n, would be beyond the power of even the fastest computers.

Before proceeding to the encipherment process itself here are two examples showing how d can be found if the values of p, q and e are known. The first example involves only small numbers; in the second the numbers are somewhat larger although still much smaller than would be likely to be used in RSA encipherment.

Example 13.1

Find the decipherment key, d, when $n=91$ and the encipherment key, e, is 29.

Solution

We begin by noting that $91 = 7 \times 13$ and so, taking $p=7$ and $q=13$ we have

$$(p-1)(q-1) = 6 \times 12 = 72$$

and since 29 has no factor in common with 91 or 72 it is a valid encipherment key for the RSA method. To find the corresponding decipherment key, d, we use the Euclidean Algorithm which is explained in M25, but the method is illustrated by the following:

$$72 = 2 \times 29 + 14,$$
$$29 = 2 \times 14 + 1,$$

so

$$1 = 29 - 2 \times 14,$$

but we have seen, in the first line above, that $14 = 72 - 2 \times 29$ and so

$$1 = 29 - 2 \times (72 - 2 \times 29) = 5 \times 29 - 2 \times 72, \text{ that is:}$$
$$5 \times 29 = 2 \times 72 + 1,$$

or

$$5 \times 29 \equiv 1 \pmod{72}$$

(i.e. $145 = 144 + 1$) which means that d, the decipherment key, is 5.

This may seem confusing, but the method is to use the Euclidean Algorithm to find the highest common factor of the encipherment key, e, and the number $(p-1)(q-1)$. Since e and $(p-1)(q-1)$ have no common factor the highest common factor is 1. If we now 'work backwards' through the Euclidean Algorithm from the last line to the first, replacing the last number on the right of each identity by the other numbers involved, we will eventually obtain the decipherment key, d, as given by the congruence in (3) above. A formal description, and an alternative approach, are described in M25.

Example 13.2 (A slightly more realistic case, which we shall use below)

Find the decipherment key, d, when $n = 3127$ and the encipherment key, e, is 17.

Solution

We note that $n = 3127 = 53 \times 59$ and 53 and 59 are both primes. It follows that n is a valid modulus for the RSA and that $(p-1)(q-1) = (52)(58) = 3016$. Furthermore since 17 has no factor in common with 3127 or with 3016 it is a valid encipherment key. Proceeding as before, that is, as in the Euclidean Algorithm:

$$3016 = 177 \times 17 + 7,$$
$$17 = 2 \times 7 + 3,$$
$$7 = 2 \times 3 + 1,$$

so

$$1=7-2\times3$$

but, from the second line above, $3=17-2\times7$ and so

$$1=7-2\times(17-2\times7)=5\times7-2\times17$$

but, from the first line above, $7=3016-177\times17$ and so

$$1=5\times(3016-177\times17)-2\times17=5\times3016-887\times17$$

or

$$887\times17\equiv-1\ (\text{mod } 3016).$$

We need a value of d that, when multiplied by 17, leaves remainder $+1$ when divided by 3016 and this is therefore -887, not $+887$. Now $-887\equiv 2129\ (\text{mod } 3016)$ so, finally, the decipherment key, d, is 2129.

(Check: $2129\times17=36\,193=12\times3016+1$.)

The encipherment and decipherment processes in the RSA system

To *encipher* a message using the RSA method the would-be user must know the values of the modulus, n ($=pq$), and the encipherment key, e; these are publicly available. Only the 'owner' of the system which is being used knows the values of p and q and the decipherment key, d.

To encipher a message which is to be sent to the 'owner', and which he alone can decipher, the user must:

(1) convert the letters of his message into numbers such as those given in Table 1.1, so that, for example A = 00, B = 01, ..., Z = 25;
(2) if the modulus n contains no more than D digits break the number version of the message into blocks of no more than D digits; denote these blocks by $B_1, B_2, \ldots$;
(3) encipher the blocks in order and independently by computing
(4) $(B_I)^e\ (\text{mod } n)$, $I=1, 2, \ldots$, giving cipher blocks $C_1, C_2, \ldots$

The cipher message is $C_1C_2\ldots$ (a string of numbers, not letters).

To *decipher* a message the procedure is exactly the same as for encipherment except that the blocks to be handled now are the *cipher* blocks C_I and the *decipher* key, d, is used instead of the encipher key, e, at stage (4). It is a consequence of the way that d has been determined that

$$(C_I)^d = B_I$$

or, in other words, that the original message is recovered.

Example 13.3

Encipher the message

COMEXATXNOON

using the RSA System with $n = 3127$ and encipherment key, e, $= 17$.

Encipherment

We convert the text into numbers using Table 1.1 in the usual way. Since the modulus, 3127, contains four digits we break the text up into pairs of letters,viz:

CO	ME	XA	TX	NO	ON
0214	1204	2300	1923	1314	1413

We now have to compute the value of each of the six four-digit numbers raised to the power 17 with respect to the modulus 3127. The computation is laborious by hand and so is only given in detail for the first number, 0214. The other values are easily obtained by a simple computer program.

Our task then is to compute $(0214)^{17}$ (mod 3127). Whenever possible in modular computations of this kind we use repeated squarings to get powers of the number and we then multiply together appropriate powers. Since $16 = 2^4$ we can get close to the required power and so we proceed as follows:

$$(214)^2 = 45\,796 = 14 \times 3127 + 2018 \equiv 2018 \pmod{3127}$$

so

$$(214)^4 \equiv (2018)^2 = 4\,072\,324 = 1302 \times 3127 + 970 \equiv 970 \pmod{3127}$$

and

$$(214)^8 \equiv (970)^2 = 940\,900 = 300 \times 3127 + 2800 \equiv 2800 \pmod{3127},$$

hence

$$(214)^{16} \equiv (2800)^2 = 7,840,000 = 2507 \times 3127 + 611$$
$$\equiv 611 \pmod{3127}.$$

Finally then

$$(214)^{17} \equiv (214)(611) = 130\,754 = 41 \times 3127 + 2547$$
$$\equiv 2547 \pmod{3127}.$$

The encipherment of 0214 is therefore 2547.

The other blocks of the message text are enciphered in the same way, that is, by raising each four-digit number to the power 17 (mod 3127). The resultant cipher text is

2547 3064 2831 0063 2027 1928.

This cannot be converted back into letters since some, in fact most, digit-pairs are greater than 25.

To *decrypt* this cipher message the recipient ('owner') would raise each four-digit number to the power of the decipherment key, 2129, (mod 3127). Since

$$2129 = 2048 + 64 + 16 + 1 = 2^{11} + 2^6 + 2^4 + 1$$

the computation would involve raising each four-digit number to the powers 2^4, 2^6 and 2^{11} by repeatedly squaring and then forming appropriate products. Using this 'repeated squaring' method raising the cipher blocks to the power 2129 involves 'only' 14 multiplications or divisions instead of over 4000 which a direct, 'brute force', calculation would require. To show how this would be done and to confirm that 2129 *is* the decipherment key in this case, here is the calculation for the decipherment of the fourth four-digit block in the cipher message above (i.e. 0063).

We compute $(63)^{2129}$ and find the remainder when we divide by 3127. From above,

$$(63)^{2129} = (63)^{2048} \times (63)^{64} \times (63)^{16} \times (63)^1$$

and we proceed to make a table of 2^nth powers of 63 up to $n = 11$ by repeated squaring. Thus $(63)^2 = 3969 = 3127 + 842$ so we put 842 in the table opposite $n = 2$. We continue in this way and the full table is shown in Table 13.1. To get $(63)^{2129}$ (mod 3127) we now multiply together the four right-hand numbers opposite $n = 0, 4, 6$ and 11:

$$(63)^{2129} \equiv (63)(523)(1500)(2822)$$

Now

$$63 \times 523 = 32\,949 = 10 \times 3127 + 1679 \equiv 1679 \pmod{3127},$$

Table 13.1

n	$N=2^n$	$(63)^N$ (mod 3127)
0	1	63
1	2	842
2	4	2262
3	8	872
4	16	523
5	32	1480
6	64	1500
7	128	1687
8	256	399
9	512	2851
10	1024	1128
11	2048	2822

and

$$1500 \times 2822 = 4\,233\,000 = 1353 \times 3127 + 2169 \equiv 2169 \pmod{3127}$$

and, since

$$1679 \times 2169 = 3\,641\,751 = 1164 \times 3127 + 1923 \equiv 1923 \pmod{3127},$$

the decipherment of 0063 is therefore 1923. This converts back to letters as TX which *is* the fourth digraph of the original message.

This computation, though tedious, can be done on a pocket calculator but in a real application of the RSA system the integers involved would be very much bigger and a computer with software capable of handling such integers would be essential. Even when a computer is being used the reduction in the number of multiplications and divisions by employing the 'repeated squaring' technique is vitally important. In a typical application of the RSA system, where the modulus is likely to be at least of the order of 10^{100}, the encipherment or decipherment key might easily be around 10^{50} and, since the blocks of numbers have to be raised to this power, 'brute force' calculation is out of the question. By using repeated squaring, on the other hand, the number of multiplications and divisions is reduced to a few hundred, which would take milliseconds at most (M26).

How does the key-owner reply to correspondents?

Having received a message enciphered on the RSA system using his public key how can the owner of the decipherment key reply? The answer is very simple: by *enciphering* his message with the (secret) *decipherment* key, d; the recipient then *deciphers* the cipher message which he has received by using the (public) *encipherment* key. This will produce the original plaintext message since, as remarked earlier, there is symmetry between the encipherment and decipherment keys in the RSA system: *each unlocks the other.*

The RSA algorithm is a particular case of a *public key cipher system* since anyone wishing to receive enciphered messages provides a key in some publication which is publicly available, such as a telephone directory. This is the *encipherment* key. Only the owner of the encipherment key has the corresponding *decipherment* key so that although anyone can send him an enciphered message he alone can decipher it. For this to be so the method of encipherment must be such that it is impossible, in a reasonable time, to find the decipherment key even when the method of encipherment and the encipherment key are known. In the case of the RSA the security depends upon the fact that it is easy to multiply two large numbers together but very difficult to find the two large numbers if we are only given their product. A function which is easy to compute but very difficult to invert, such as the RSA, is called *a one-way function*. Another example is the discrete logarithm problem that we met earlier in the Diffie–Hellman system.

Problem 13.1

It is possible that in the RSA system a string will encipher to itself. The most obvious cases in the two-digit example above are the strings `AA` and `AB` which are represented numerically by 0000 and 0001 respectively and, since both of these raised to any power remain unaltered, will encipher to themselves regardless of the encipherment key or modulus. There are however other cases; verify that

> with modulus 3127 and encipherment key 17 the numbers 0530 and 0531 both encipher to themselves in the RSA system.

(Lest it be argued that 0530 and 0531 do not correspond to letter pairs let it be noted that 0825 and 2302, the numerical forms of `IZ` and `XC`, also encipher to themselves.)

It must be said, however, that the probability that a block of text in a 'real' application of the RSA would encipher to itself is negligible.

The Data Encryption Standard (DES)

In 1973 the US Government, in response to repeated requests from industry and various organisations, gave its Department of Commerce the task of establishing uniform Federal Automatic Data Processing Standards and, within that Department, the responsibility was passed to the National Bureau of Standards (NBS). One particular aspect considered by the NBS was that of producing a standard for the encryption of data. Rather than simply proposing an algorithm themselves the NBS published an invitation to interested parties to submit proposals for 'The Data Encryption Standard' in May 1973.

The specification for the Data Encryption Standard published by the NBS laid down conditions that any proposed algorithm must satisfy: that it must provide a high level of security, that the security must not be based on the secrecy of the algorithm, that it must be economical to implement electronically, efficient to use and available to all users and suppliers.

The initial response having proved to be disappointing, a second invitation was published in August 1974 as a result of which the proposal submitted by IBM was selected for evaluation in March 1975. After about 18 months of discussion and comment the IBM proposal was accepted and became 'The Data Encryption Standard', DES for short, in November 1976 [13.3, 13.4].

The DES has been the subject of a great deal of analysis and there is a considerable literature devoted to it. The full details, which are mainly of interest to specialists, can be found in numerous books including [13.5] and [13.6]. The essentials can be summarised as follows.

Background

(1) The algorithm is designed to encipher blocks of 64 bits of data under the control of a 64-bit key (K). The DES is therefore an example of what is known as *a block cipher*.
(2) Two people who wish to communicate using DES must agree on the (secret) key, K. Everything else about the DES is public knowledge.
(3) For the secret key, K, the users choose *seven* 8-bit characters (i.e. 56 bits in all) and the DES then adjoins a further 8 parity check bits to give the 64-bit secret key required.

The encipherment procedure

(4) The 64-bit block of data is subjected to an initial permutation (IP).
(5) The 64 bits of data are split into two 32-bit segments, left (L) and right (R).

(6) Forty-eight bits of the key, K, are combined with an 'expanded' 48-bit version of R in a non-linear way (the 'expansion' consists of repeating 16 of the 32 bits of R) and these 48 bits are then 'reduced' to a 32-bit string, X (say).

(7) L is replaced by R and R is replaced by the (mod 2) sum of X and L to give a new 32-bit R.

(8) Steps (6) and (7) are repeated 16 times using different 48-bit segments of K at step (6) each time.

(9) The 64 bits of the final (16th) stage are subjected to the *inverse* of the initial permutation, i.e. to $(\mathrm{IP})^{-1}$.

(10) The resulting 64 bits are the cipher.

The decipherment procedure

(11) Decipherment is carried out by using the encipherment procedure in the reverse order with the same key, K.

Security of the DES

It is worth noting that two people who wish to communicate using the DES need to agree upon the common key and this can be agreed between them by using the Diffie–Hellman key exchange system. Unless a third party can intercept and change their communications this should be secure.

As to the DES itself: many statistical, and other, tests have been made on data encrypted with various keys using the DES and have been found to be satisfactory. One particularly important one is what is known as the *avalanche* test: if one of the 64 bits of the input is changed, how many of the 64 bits of the output are changed? In a weak cipher system the answer would be '1'; in a perfect cipher system it would be 'about 32' and this is what happens in the DES. (For examples see [13.5].)

Nevertheless, even before the DES was brought into use there were arguments as to its level of security. Since the users specify a key of 56 bits there are 'only' 2^{56} possible keys for an adversary to try. Now 2^{56} is approximately equal to $10^{16.86}$ and if a computer could test one key in a microsecond it could try them all in about 2300 years. This is clearly impractical but the critics of the DES have suggested that a million such computers working in parallel could find the key in under a day. Since this would cost a huge amount of money, and take a great deal of organising, it is hard to imagine that anyone would think it worthwhile unless the message were known to relate to a vital matter of national security. It has also been suggested that perhaps there is a secret 'trapdoor', or Achilles heel, known to the designers of the DES that would enable those who know it to find the key in a realistic time. This could be true, but nobody has found any evidence that such a thing exists.

It has also been suggested that the DES should have been based upon a longer key, 128 bits being the most popular choice. This would certainly put its security beyond question and, since then, 128-bit encipherment has been introduced in other systems, but in 1977 it was not thought to be necessary.

An alternative to using 128-bit keys would seem to be using two 64-bit keys in succession, but this can be shown to be only twice as secure as using a single 64-bit key provided that a vast amount of computer memory is available. The attack assumes that a known plaintext and its cipher equivalent are available. The known plaintext is *enciphered* under all possible 64-bit keys and the known cipher text is independently *deciphered* under all possible 64-bit keys. The two sets of data are sorted and compared. This means that 2^{57} tests are required rather than 2^{112}. When two identical texts are found we have candidates for the two unknown keys. Many false key pairs (about 2^{48} in fact; see M27) will be found and will have to be tested on further known plaintext–cipher text pairs. This 'meet-in-the-middle' attack is not practical at present and is unlikely to be so in the foreseeable future.

The security of the DES is enormously increased however if we use *triple* encipherment. Only two keys are required but they are employed as follows:

(1) *encipher* with key 1;
(2) *decipher* with key 2;
(3) *encipher* with key 1.

The adversary now will have to try 2^{112} possible pairs of keys and this is considered to be impossible in any realistic time. This *triple DES encipherment* is regarded as secure and is currently in use. This form of triple encipherment has the additional advantage that it becomes identical to single encipherment DES if the two keys are the same, for it would allow a user with triple encipherment to communicate with a single encipherment user, and vice versa. Another form of triple encipherment uses *three* different keys in the three stages above. Once again compatibility with the other forms is achieved by making two or more of the keys the same.

Chaining

Since the DES enciphers text in short blocks of only 64 bits the obvious question is: how do we encipher messages that are longer than this? The simplest way is to break the message into blocks of 8 characters (=64 bits) and encipher them sequentially using the same 64-bit key. This would mean that all the 64-bit cipher messages were 'in depth', but the non-linear nature of DES encipherment makes this feature, which would be disastrous if the encipherment were linear, of little or no value to the cryptanalyst. A more secure method however is to change the key for each 8-character block by making each key depend on the original key and some, or all, of the *plaintext* of the preceding blocks. An authorised recipient will recover the plaintext of the first block since he knows the original key; he will therefore be able to construct the key for the second block and so decipher it. He will now have the plaintext of the second block and so be able to construct the key for deciphering the third block; and so on. An unauthorised recipient who manages to break into part of the message, possibly because of some repetitive standard text, would not be able to progress further because, without knowing *all* of the earlier blocks of plaintext, he cannot reconstruct the other keys. Had the same key been used for each block he would have been able to decrypt the entire message.

Users of encipherment systems that are based on keys applied to short blocks of text, such as the DES, are strongly recommended to use chaining.

Implementation of the DES

Although it is not difficult to write a program to encipher/decipher using the DES algorithm no software implementation can be approved, partly because programs can be modified. In addition, software versions would be much slower than hardware versions on specifically designed chips and, shortly after the approval of the DES, various manufacturers designed and produced devices which contained chips for carrying out the DES algorithm. These devices can encipher or decipher at rates of a hundred thousand characters and more, per second.

Using both RSA and DES

In public key systems, such as RSA, the encipher/decipher algorithm generally involves a great deal of computation and so may run 'slowly', whereas in block cipher systems, such as DES, the encipher/decipher processes can be carried out much faster, perhaps a thousand times faster in fact. If however

(1) the sender encrypts the DES key using the RSA method and sends it to the recipient, and then
(2) the sender encrypts the message with the DES key,

he will keep the amount of computation down considerably without reducing the security as a whole. If 'chaining' is being employed the RSA algorithm would only have to be used once, to encipher the initial DES key.

A salutary note

From a 'brute force' point of view simple substitution ciphers, where there are more than 10^{26} possibilities to consider, should be more difficult to solve than DES encrypted messages where the number of possibilities is less than 10^{17}. This once again illustrates how misleading 'brute force' arguments can be.

Beyond the DES

The original DES was given official US Government Approval for about 10 years. After that time triple DES, as described above, continued to be approved but, in addition, new encryption algorithms were sought. In 1993 an algorithm called Skipjack was authorised and implemented on a chip known as Clipper. The details of the Skipjack algorithm were initially kept secret but were declassified on June 24th 1998 [13.7].

An 'unofficial' encryption system (i.e. one that was developed by private individuals) that has attracted many users is PGP, which is short for 'Pretty Good Privacy'. PGP was developed by Philip Zimmermann and is freely available to anyone who wants it. It involves the following steps:

(1) a 'random' key is generated based upon the user's movements of the 'mouse' and keystrokes; this is referred to as the *session key*;
(2) the plaintext of the message is compressed; the extent of compression depends upon the nature and length of the text; a typical English text might be compressed by 50% without introducing ambiguities but for short texts the saving, in transmission time, may not be worthwhile;
(3) the key generated in (1) is used to encipher the compressed text produced in (2) using an algorithm called IDEA ('International Data Encryption Algorithm') invented by Lai and Massey at ETH, Zurich [13.8];

(4) the session key is now encrypted using the public key of the recipient;
(5) the encrypted session key is placed at the front of the encrypted text and the whole sent to the recipient;
(6) on receiving the encrypted text the recipient first uses his private key to decrypt the session key at the beginning of the text;
(7) with this decrypted key he can now decrypt the message text, and decompress it if necessary.

As in the case of RSA followed by DES mentioned above, the use of two encryption systems in PGP not only increases security, it also considerably increases speed because the time taken to encrypt or decrypt using a system where the encryption/decryption keys are essentially the same is far less than the time required to encrypt/decrypt in a public key system, such as RSA, where the keys are totally different. In PGP only the relatively short session key is encrypted by a public key system; the text itself is encrypted by a non-public key system (IDEA).

PGP and related topics have led to thousands of articles that can be found on the internet. Anyone who wants more details should consult [13.10], [13.11].

Encryption algorithms for public use have been the subject of a great deal of research and it is to be expected that this will continue.

Authentication and signature verification

As was mentioned earlier, these problems can be solved using public key systems. The DES is not such a system but the RSA method is. Recall that in a public key system each user has a public (encipherment) key, E, and a secret (decipherment) key D. Let us suppose that X wishes to send a message, M, to Y and that

$$E_X, D_X, E_Y \text{ and } D_Y$$

are the encipherment and decipherment keys of X and Y respectively.

How can:

(1) Y be sure that the message *has* been sent by X?
(2) X ensure that Y cannot claim to have received a different message, M', say?
(3) Y ensure that X cannot claim that he sent a different message, M'', say?

There are several ways of solving these problems, all of them involve using some or all of the encipherment and decipherment keys. A typical solution is:

(1) X precedes M with information which includes the date and time as well as his identity, thus M is extended to a longer message $M1$, say, which is something like

$M1$: 'I am X, the date is 13-06-2001 and the time is 1827. M'

where M is the original message.

(2) X *decrypts* $M1$ using his private key, D_X, producing a cipher message $D_X(M1)$ which he sends to Y.

(3) Y applies X's public *encipherment* key, E_X, to this cipher message and so recovers the extended message, $M1$, since

$E_X(D_X(M1)) = M1$.

Now:

(1) Y can be sure that X sent $M1$ since only X can produce $D_X(M1)$.

(2) If Y claims that he received a different message, M', X challenges Y to produce the cipher text $D_X(M')$. Y will be unable to do this since he doesn't know X's secret key, D_X.

(3) If X claims that he sent a different message, M'', to Y the latter can show the cipher message, $D_X(M1)$, to a judge who will ask X for his secret key so that he can check whether the message $M1$ *was* sent. Since Y doesn't know X's secret key he couldn't have constructed $D_X(M1)$. If X refuses to give his secret key to the judge he will lose his case.

It is important that the extended message $M1$ includes the date and time otherwise an old message could be substituted for $M1$ which would invalidate the point made in (2) above.

Elliptic curve cryptography

In recent years an interesting method for signature verification has received a lot of attention both in the Universities and in industry. The method, known as *Elliptic Curve Cryptography* (ECC for short), is based upon deeper mathematical ideas than the RSA algorithm and is claimed to be more secure. The mathematics behind the method is too advanced to be described here but interested readers are invited to turn to M28.

Appendix
Mathematical aspects

Chapter 2

M1 Identical letters in substitution alphabets

This, the problem of the number of *derangements* of the elements of a set, follows as a special case of what is known both as *the classical sieve formula* and as *the inclusion and exclusion principle*, proofs of which can be found in books on combinatorics such as [2.6]. The sieve formula tells us that the fraction of the number of permutations of the numbers $(1, 2, \ldots, n)$ which are such that no number is in its 'correct' position, i.e. of derangements of n numbers, is

$$\frac{1}{1} - \frac{1}{1!} + \frac{1}{2!} - \frac{1}{3!} + \frac{1}{4!} - \frac{1}{5!} + \cdots \text{ for } (n+1) \text{ terms}$$

and this fraction rapidly converges to the value 0.3678... which is the reciprocal of the number e, for those familiar with *natural logarithms*. For $n=0$ to 6 the fractions have the values to three decimal places 1, 0, 0.5, 0.333, 0.375, 0.367, and 0.368. Thus the fraction is virtually the same in practical terms for values of n greater than 5. This means that a permutation alphabet on 26 letters has approximately a 37% chance of having no letter in its 'correct' place and hence a 63% chance of having at least one letter in its original position.

M2 Reciprocal alphabets weaken security

We can choose the first pair of letters in

$$\frac{26 \times 25}{2}$$

ways, since choosing to pair, say, A and W is the same as choosing to pair W and A. Similarly the second pair can be chosen in

$$\frac{24 \times 23}{2}$$

ways, and so on. There would therefore seem to be

$$\frac{26 \times 25 \times 24 \times 23 \times \cdots \times 4 \times 3 \times 2 \times 1}{2 \times 2 \times \cdots \times 2 \times 2}$$

ways of forming reciprocal alphabets but this is not so, for we can re-arrange the 13 pairs in any order without changing the substitution alphabet. For example, if we choose to pair A and W and then to pair B and K we would get exactly the same result as if we first paired B and K and then paired A and W. We must therefore reduce the number above by the factor

$$13! = 13 \times 12 \times 11 \times \cdots \times 2 \times 1$$

which is greater than 6227 000 000 and since the 13 factors of 2 in the denominator above provide a further factor of 8192 we see that we have reduced the number of substitution alphabets by a factor of more than 50 000 000 000 000. Overall this means that the number of possible substitution alphabets is reduced from more than 10 to the 26th power to fewer than 10 to the 13th power.

It may seem strange, but it is better not to pair *all* of the 26 letters; the number of possibilities is increased if only 22 are paired and the other 4 letters left unchanged. This is because the number of possibilities if we pair $2k$ letters and leave $(26-2k)$ unchanged is

$$\frac{(26!)}{(k!)((2n-2k)!)2^k}$$

and this reaches a maximum at $k=11$. Whilst this fact is not important for simple substitution ciphers it *is* of significance in the context of the number of pairings on the Enigma plugboards, as we shall see in Chapter 9.

M3 The birthdays paradox

The probability that two people chosen at random have the same birthday is 1/365. We ignore leap years, which have no significant effect on the result.

Suppose that we have already looked at n people and that no two of them have the same birthday. Then when we look at the $(n+1)$st person the probability that he/she will not have a birthday in common with any of the others is

$$\frac{(365-n)}{365}$$

The probability that no pair of people among 23 people chosen at random will have the same birthday is therefore

$$\frac{364}{365}\times\frac{363}{365}\times\frac{362}{365}\times\cdots\times\frac{343}{365}$$

the value of which (to three decimal places) is 0.493. Therefore the probability that *at least* one pair will have the same birthday is $(1-0.493)$, which is 0.507, and since this number is greater than a half there is a better than evens chance of there being a pair with the same birthday. Had we confined our attention to 22 people, rather than 23, the probability of there being at least one pair with the same birthday would have been less than a half, 0.476 to three decimal places.

Chapter 3

M4 Euclid's proof that there are an infinite number of primes

Suppose, on the contrary, that there are only a finite number of primes and that they are

$$2, 3, 5, 7, 11, \ldots, P,$$

Consider the number, N, formed by multiplying all of them together and adding 1:

$$N=2\times3\times5\times7\times11\times\cdots\times P+1.$$

Clearly N is not divisible by 2 or 3 or 5 or 7 or 11 or ... or P since it leaves remainder 1 on each such division. So N is not divisible by any prime in the list. It is therefore either itself a prime or divisible by some prime which is not in the list, and in either case our alleged list of all the primes must be incomplete. There are therefore an infinite number of primes.

For example; if someone claims that the only primes are 2, 3 and 5 then $N=31$, which is a prime. If he then adds 31 to his list of primes then

$$N=2\times 3\times 5\times 31+1=931=7\times 7\times 19$$

and 7 and 19 are primes which are not in his list, and so on, for ever.

Chapter 6

M5 The Fibonacci sequence

If U_n denotes the nth term in the sequence then the sequence is generated by the *linear recurrence*

$$U_{(n+1)}=U_n+U_{(n-1)} \quad \text{where } U_0=0 \text{ and } U_1=1.$$

The standard method for finding the general solution of any second order linear recurrence such as this is to assume that it takes the form

$$U_n=A\{\alpha^n\}+B\{\beta^n\}$$

where A, B, α and β are constants. Substituting this expression into the recurrence we find that this is a valid assumption if α and β are the roots of the quadratic equation

$$X^2-X-1=0.$$

Taking α as the positive root we have

$$\alpha=\frac{1+\surd 5}{2} \text{ and } \beta=\frac{1-\surd 5}{2}$$

or, numerically, $\alpha=1.6180...$ and $\beta=-0.6180...$

The values of A and B are then found by imposing the conditions that $U_0=0$ and $U_1=1$ which give

$$A=-B=\frac{1}{\surd 5}.$$

For large values of n the value of U_n is the integer nearest to $A\alpha^n$ so that each term is approximately 1.6180... times its predecessor. Thus the 8th term is the integer nearest to

$$\frac{(1.6180)^8}{\surd 5}.$$

The value of this to three decimal places is 21.006; the nearest integer is therefore 21 and the 8th term of the Fibonacci sequence is indeed 21.

Discussion of the Fibonacci sequence will be found in many books on elementary number theory. The sequence has a long history. Fibonacci, also known as Leonardo of Pisa, introduced it in his book, *Liber Abaci*, in 1207. The sequence also has a very large number of properties; for example, every 5th term is divisible by 5, every 8th term is divisible by 7 and every 10th term is divisible by 11. Such properties, though very nice from a mathematical point of view, make the sequence quite unsuitable cryptographically. For an extensive study of the sequence see [6.5]. There is also a journal devoted to the study of the Fibonacci and other linear sequences [6.6]. Related material will also be found in articles on the topic of *continued fractions* [6.7].

Chapter 7

M6 Letter frequencies in a book cipher

In a book cipher where 'space' and punctuation symbols are collectively regarded as a 27th letter a cipher letter can appear as a result of 27 combinations of a key letter added to a message letter. Thus, for example, in order to produce the letter D in the cipher we need one of the following 27 combinations:

	A in the key and D in the message
or	B in the key and C in the message
or	C in the key and B in the message
or	D in the key and A in the message
or	E in the key and 'space' in the message
	etc.
or	Z in the key and F in the message
or	'space' in the key and E in the message.

If we let $p(\#)$ denote the probability of a particular letter, #, occurring in an English text then the probability of the letter D appearing in a book cipher text is

$$p(\mathrm{A})p(\mathrm{D})+p(\mathrm{B})p(\mathrm{C})+p(\mathrm{C})p(\mathrm{B})+p(\mathrm{D})p(\mathrm{A})+\cdots+p(\text{'space'})p(\mathrm{E}).$$

Using the table of typical frequencies of the letters in English texts, counting all punctuation and 'space' as a 27th letter, we can compute the expected probability of D occurring in a book cipher from this expression, and similarly for any other letter.

A particular case relates to the probability of two messages being in depth, mentioned in Chapter 3. If we draw a pair of letters at random from two plaintext messages, where only the letters A to Z occur, punctuation being omitted, the probability that the two letters will be the same is

$$p(\mathrm{A})^2 + p(\mathrm{B})^2 + \cdots + p(\mathrm{Z})^2$$

and this turns out to be about 1/13 for English and, more generally, for most natural languages, about

$$\frac{2}{(\text{alphabet size})}.$$

Thus, for example, taking the frequencies of the 27 letters (alphabet + punctuation symbol) in Table 7.4 we see that the probability of two letters being the same in this case is

$$\frac{(64)^2 + (14)^2 + \cdots + (166)^2}{1\,000\,000} = \frac{70\,678}{1\,000\,000} \approx \frac{1}{14}.$$

M7 One-time pad cipher cannot be solved

In a one-time pad all the letters occur with equal frequency and so, no matter how many letters of the key we have seen, we cannot predict what the next letter will be. Thus all keys are equally likely and this means that a one-time pad cipher message could be 'deciphered' to produce *any* plaintext message of the appropriate length since the alleged key has no properties to distinguish it from any other. For example if the cipher message

QLXEB YEMUC AFNQQ

has been enciphered using a one-time pad, then if the key from the pad had been

JLIPD BCFDU IMBQY

the decrypt would be

HAPPYXCHRISTMAS.

If, however, the key had been

DRVTX YNPIU INFFM

the decrypt would be

NUCLEARXMISSILE,

and, since all random keys of 15 letters are equally likely, either of these could be correct and indeed there are more than 10^{21} other possible decrypts, most of which are, however, nonsense.

Chapter 8

M8 Frequency of occurrence in a page of random numbers

In a page of 100 two-digit random numbers any number in the range 00 to 99 can be expected to occur once. The probability that a particular number will not occur in any particular position is 0.99 and, since the numbers are random the probability that any particular number will not occur at all in a page of 100 is

$$(0.99)^{100} \text{ or } (1-1/100)^{100}$$

the value of which is, effectively, e^{-1}, where $e = 2.718\,28...$ is the base of natural logarithms, as was remarked before, in M1. Since $e^{-1} = 0.37$ to 2 d.p. it follows that in a typical page of 100 random two-digit numbers there will be about 37 that do not occur. On the other hand there should be about

$$\frac{100e^{-1}}{3!}$$

which occur three times, i.e. about 6, and we might expect one number to occur four times since the expected number in that case is

$$\frac{100e^{-1}}{4!}$$

the value of which lies between 1 and 2.

(What we are doing, in effect, is claiming that the probability of a specified two-digit number occurring exactly n times on a page of 100 such random numbers is approximately

$$\frac{e^{-1}}{n!}$$

which is a particular case of what is known as *the Poisson distribution* in probability theory. For a full mathematical treatment and justification, consult books on probability theory, such as [8.1].)

M9 Combining two biased streams of binary key

If the streams are unrelated but have a bias towards 0 the probability of which is $(0.5+x)$ then if we form the (mod 2) sum of the streams the probability of 0 will be

$$(0.5+x)^2+(0.5-x)^2=0.5+2x^2.$$

So, for example, if $x=0.01$ the bias in the combined stream will be only 0.0002. (If the two streams are in any way related this argument is false as is obvious since if, for example, the two streams are identical their (mod 2) sum consists of all 0s.)

M10 Fibonacci type sequence

Following the notation of M5 the terms of this sequence are generated by the linear recurrence

$$U_{(n+1)}=2U_n+U_{(n-1)}.$$

To prove that every third term is divisible by 5 we note that the recurrence formula gives

$$U_n=2U_{(n-1)}+U_{(n-2)}$$

and so, substituting for U_n in the expression for $U_{(n+1)}$, we obtain

$$U_{(n+1)}=5U_{(n-1)}+2U_{(n-2)}.$$

If $U_{(n-2)}$ is divisible by 5 it follows that $U_{(n+1)}$ must also be divisible by 5. Since U_0, which has the value 0, is divisible by 5 then so are U_3, U_6 and so on.

As to the ratio of consecutive terms approaching the value $(1+\sqrt{2})$: as before, we assume that U_n may be represented in the form

$$U_n=A(\alpha^n+\beta^n).$$

Applying the recurrence formula and the initial conditions that $U_0=0$ and $U_1=1$ we find that our assumption is valid if α and β are the roots of the quadratic equation

$$X^2-2X-1=0$$

and $B=-A$, which lead to the values

$$\alpha=(1+\sqrt{2}),\ \beta=(1-\sqrt{2}) \text{ and } A=1/(2\sqrt{2}).$$

The ratio of consecutive terms therefore rapidly approaches the value of α, for β is less than 1 in absolute value and so the values of its powers very quickly become small. Since $\alpha = (1+\sqrt{2})$ this proves the assertion.

M11 Binary linear recurrences

As might be expected after seeing the analysis of the Fibonacci and related sequences we need to investigate the properties of the polynomial of degree k that is related to the linear recurrence of order k. Such a recurrence may be written

$$U_n = a_1 U_{(n-1)} + a_2 U_{(n-2)} + \dots + a_k U_{(n-k)}$$

and the associated polynomial is

$$X^k + a_1 X^{(k-1)} + \dots + a_k = 0.$$

Remember that we are working (mod 2) so that $-a_s$ is the same as $+a_s$. Since the recurrence is a binary recurrence and is of order k all the coefficients (or *multipliers*) are 0 or 1 and furthermore we can take the last coefficient, a_k, to be 1. There are therefore $2^{(k-1)}$ possibilities for the coefficients.

The mathematical analysis of the periods of binary sequences generated by a linear recurrence is too deep to go into here but the key result, so far as we are concerned, can be stated as follows.

Theorem

Let $\phi(n)$ denote the number of integers which are less than n and which have no factors in common with it. Then the number of binary linear recurrences of order k which generate a key stream of maximum length $(2^k - 1)$ is

$$\frac{\phi(2^k - 1)}{k}$$

The function $\phi(n)$ is known as Euler's ϕ (pronounced 'phi') function. It is easily computed. Let $p_1, p_2, \dots, p_r$ be the different primes *any* power of which exactly divides n. Then

$$\phi(n) = \frac{n(p_1 - 1)(p_2 - 1)\dots(p_r - 1)}{p_1 \qquad p_2 \qquad\qquad p_r}$$

Thus, for example, taking $k = 12$ we have

$$2^{12} - 1 = 4095 = 3 \times 3 \times 5 \times 7 \times 13$$

and so

$$\phi(4095) = 4095 \times \frac{2 \times 4 \times 6 \times 12}{3 \times 5 \times 7 \times 13} = 1728$$

and the number of binary linear recurrences of order 12 which generate binary sequences of maximum length 4095 is therefore

$$\frac{1728}{12} = 144$$

as stated in Chapter 8. Note that although 4095 is divisible by 3^2 the fraction in the Euler function is two-thirds, *not* eight-ninths.

Similarly, when $k = 23$ the number of binary linear recurrences of order 23 which generate binary sequences of maximal length $(2^{23} - 1)$ is

$$\frac{\phi(2^{23} - 1)}{23}$$

which is

$$\frac{\phi(47 \times 178\,481)}{23}$$

and 47 and 178 481 are primes (that 178 481 *is* a prime follows from the fact that it is not divisible by any prime less than its square root, i.e. by any prime less than 422). We therefore find that the number of such sequences is

$$\frac{46 \times 178\,480}{23} = 356\,960$$

which was also mentioned in Chapter 8.

(For a proof of the theorem quoted above see [8.2].)

M12 Recovery of a binary linear recurrence from a stretch of key

Example

The following 15 binary digits of key have been recovered from a cipher message:

10101 00110 00100

the most recent digits being to the right. Verify that these can be generated by a linear recurrence of order 5.

Solution

A linear recurrence of order 5 takes the form

$$U_n = aU_{(n-1)} + bU_{(n-2)} + cU_{(n-3)} + dU_{(n-4)} + eU_{(n-5)}$$

where a, b, c, d and e are unknown constants whose values are either 0 or 1, since all the arithmetic is (mod 2).

We number the bits 1 to 15 from left to right and then put $n = 6, 7, 8, 9$ and 10 in the recurrence to obtain five linear equations in the five unknowns a, b, c, d and e:

$$a(1) + b(0) + c(1) + d(0) + e(1) = 0, \quad \text{(A.1)}$$

$$a(0) + b(1) + c(0) + d(1) + e(0) = 0, \quad \text{(A.2)}$$

$$a(0) + b(0) + c(1) + d(0) + e(1) = 1, \quad \text{(A.3)}$$

$$a(1) + b(0) + c(0) + d(1) + e(0) = 1, \quad \text{(A.4)}$$

$$a(1) + b(1) + c(0) + d(0) + e(1) = 0. \quad \text{(A.5)}$$

From equations (A.1) and (A.3) we find that $a = 1$ and then from equation (A.4) it follows that $d = 0$. From equation (A.2) we then find that $b = 0$ and from equation (A.5) that $e = 1$ and, finally, from equation (A.1) that $c = 0$. The solution to these five equations is therefore

$$U_n = U_{(n-1)} + U_{(n-5)}.$$

We now need to confirm that this gives the correct values when $n = 11, 12,$ 13, 14 and 15 and it will be seen that this is the case. We have therefore verified that the recurrence of order 5 that we have just found *does* generate the given stretch of key.

As mentioned in Chapter 8 it can happen that there are *no* solutions of order k or that there is more than one solution. In the former case the equations are inconsistent and in the latter the ambiguities can usually be resolved if extra key digits are available. The following examples illustrate these situations.

Example (More than one solution)

Verify that the 10-bit binary key

0110110110

can be generated by either of two binary linear recurrences of order 5.

Verification

Let the recurrence be

$$U_n = aU_{(n-1)} + bU_{(n-2)} + cU_{(n-3)} + dU_{(n-4)} + eU_{(n-5)}.$$

We number the bits 1 to 10 from left to right and put $n = 6, 7, 8, 9$ and 10 successively in the recurrence which yields the equations

$$\begin{aligned}
a \quad + c + d \quad &= 1,\\
a + b \quad + d + e &= 0,\\
b + c \quad + e &= 1,\\
a \quad + c + d \quad &= 1,\\
a + b \quad + d + e &= 0,
\end{aligned}$$

which have two genuine sets of solutions of order 5 (that is, where e, the coefficient of $U_{(n-5)}$, is not zero):

$$a = b = c = 0, d = e = 1$$

and

$$a = d = 0, b = c = e = 1.$$

The two linear recurrences of order 5 are therefore

$$U_n = U_{(n-4)} + U_{(n-5)}$$

and

$$U_n = U_{(n-2)} + U_{(n-3)} + U_{(n-5)}.$$

In addition, there are solutions where $e = 0$, that is, solutions which are not of order 5, these include

$$a = b = 1, c = d = e = 0$$

corresponding to the recurrence, which is of order 2,

$$U_n = U_{(n-1)} + U_{(n-2)},$$

which reveals the fact that the binary sequence above is just that of the Fibonacci numbers (mod 2).

Example (No solution having the assumed order)

Verify that the six-bit binary key

011010

cannot be generated by any binary linear recurrence of order 3.

Verification

If such a recurrence exists it takes the form

$$U_n = aU_{(n-1)} + bU_{(n-2)} + cU_{(n-3)}.$$

The data then give the equations

$$\begin{aligned} a + b \quad &= 0, \\ b + c &= 1, \\ a \quad + c &= 0. \end{aligned}$$

Adding the first and second equations (mod 2) we get

$$a + c = 1$$

which contradicts the third equation. The equations are therefore inconsistent and no solution of order 3 exists.

M13 Generation of pseudo-random numbers

In a typical application, such as where we require random numbers which are uniformly distributed over the interval [0, 1], the integers generated by the recurrence are divided by the modulus. The 16 integers in Example 8.4 would thus be divided by 17 to give the following 16 pseudo-random numbers (to two decimal places):

0.29, 0.12, 0.59, 0.00, 0.24, 0.94, 0.06, 0.41, 0.47, 0.65, 0.18, 0.76, 0.53, 0.82, 0.71, 0.35.

In realistic-sized applications it would be necessary to use a very large modulus and, even better, to utilise more than one linear recurrence and then combine the results in some way, to provide a less predictable set of values. For a useful discussion of these matters, suggested sets of values for the modulus, multiplier and increment, as well as relevant computer programs, see Chapter 7 of [8.4].

Chapter 9

M14 Wheel wirings in the Enigma

The simple substitution alphabets provided by an Enigma wired wheel at each of the 26 positions of that wheel can be represented by a 26×26 matrix. The first *column* of the matrix shows the encipherment of the 26 letters at setting 1 of the wheel, the second column shows the encipherment of the letters at setting 2 of the wheel, and so on. Because of the 'diagonal property' the entire matrix is determined as soon as the first column has been written down.

Also, the first *row* of the matrix shows the encipherment of the letter A at each of the 26 settings of the wheel and again, because of its 'diagonal property', the matrix is completely determined as soon as the first row has been written down.

The *columns* of the matrix cannot contain any repeated letters since two different letters cannot encipher to the same letter at the same wheel setting. The *rows* of the matrix may however contain repeated letters since there is no reason why a letter should not encipher to the same letter at two, or more, settings of the wheel. From a cryptographic viewpoint it would be nice if each letter enciphered to a different letter at each of the 26 wheel settings. Unfortunately, with a wheel containing 26 wires, this is impossible, as we now see.

Instead of dealing with letters we shall use numbers, which makes the argument more apparent. The 26×26 matrix must have, as its first column, a permutation of the numbers $(0, 1, 2, \ldots, 25)$; let this be

$$(a_1, a_2, \ldots, a_{25}, a_{26}).$$

Since this is a permutation of $(0, 1, 2, \ldots, 25)$ it follows that the sum

$$(a_1 + a_2 + \cdots + a_{25} + a_{26}) = (0 + 1 + 2 + \cdots + 24 + 25) = 325 \equiv 13 \pmod{26}.$$

On the other hand the numbers in the top *row* of the matrix are (mod 26)

$$(a_1, a_{26} + 1, a_{25} + 2, \ldots, a_3 + 24, a_2 + 25)$$

and, if there are no repeats among these 26 numbers, we must have that these also sum to a number which leaves remainder 13 when divided by 26, but the sum of these numbers is

$$(a_1 + a_2 + \cdots + a_{25} + a_{26}) + (1 + 2 + \cdots + 24 + 25)$$

and, since $(1 + 2 + \cdots + 24 + 25) = 325 \equiv 13 \pmod{26}$, this is (mod 26)

$$13 + 13 = 26 \equiv 0 \pmod{26}.$$

We therefore have a contradiction and it follows that the matrix rows must contain at least one repeated number.

The same argument shows that any wheel with an *even* number of wires must give rise to plain–cipher repeats, but the argument fails if the wheel has an *odd* number of wires and then we can find wheels which produce no repeats, thus:

Example (No repeats in any row of the encipherment matrix)
Consider the 7-point wheel having encipherment matrix with first *column* $(1, 3, 6, 2, 0, 5, 4)$. The first *row* is then

$$(1, 4+1, 5+2, 0+3, 2+4, 6+5, 3+6)$$

which is (mod 7)

$$(1, 5, 0, 3, 6, 4, 2),$$

a set which contains no repeats. Since the entire matrix is determined by any row or column there can be no repeats in any row.

M15 Number of possible Enigma reflectors
In the reflector the 26 letters are joined in pairs. The first pair can be chosen in

$$\frac{26\times 25}{2}$$

ways (we must divide by 2 because it doesn't make any difference which of the pair we choose first and which second). We can now choose the next pair in

$$\frac{24\times 23}{2}$$

ways; and so on. Thus we can choose to join the 26 letters into pairs in

$$\frac{26!}{2^{13}}$$

ways. We would however get the same reflector if we chose the same 13 pairings in a different order and since we can re-order the 13 pairs in 13! ways the total number of *distinct* reflectors is

$$\frac{26!}{2^{13}\times 13!}$$

and this is more than 7×10^{12}; this is the same as the number of reciprocal simple substitutions [M2].

The same calculation applies to the number of possible plugboards.

M16 Probability of a 'depth' in Enigma messages

With N messages the number of pairs of indicators is

$$\frac{N(N-1)}{2}$$

and since there are 17 576 possible starting positions for the three wheels the number of pairs of indicators that would be expected to be the same, at random, is

$$\frac{N(N-1)}{35152}$$

If this number is greater than 1 we would expect at least one repeat at random. Since the number *is* greater than 1 when N is 188 or more we see that 200 messages would certainly be more likely than not to produce a repeated indicator.

M17 Expected number of indicators needed to obtain full chains

This problem is a particular case ($N=26$) of the following which, with a number of variants, under the name of 'the coupon collector's problem' or 'the cigarette card problem', has attracted a great deal of attention over many years:

> There are *N* different items in a set. There is a very large stock of items available. A collector obtains one item at a time, at random, from the stock. How many items might he expect to have to obtain before he has a complete set?

(For those familiar with the phraseology of probability theory this is a case of *sampling with replacement*.)

It can be shown (e.g. see [8.1], p. 225) that the expected number is

$$N\left(1+\frac{1}{2}+\frac{1}{3}+\cdots+\frac{1}{N}\right)$$

and this can be estimated by replacing the sum inside the brackets by the corresponding integral

$$\int_1^N \frac{dx}{x} = \ln(N)$$

where ln(N) is the natural logarithm of N. More precisely:

$$1+\frac{1}{2}+\frac{1}{3}+\cdots+\frac{1}{N} \to \ln(N)+\gamma \text{ as } N\to\infty$$

where γ, which is known as *Euler's constant*, has the value 0.577....

Putting $N = 26$ we obtain the estimate

$$26(\ln(26) + 0.577) \approx 99.7$$

for the number of messages the cryptanalyst is likely to need before the chains can be completely determined.

Chapter 10

M18 Number of possible Hagelin cages

The number is given by

> the number of ways of representing 27 as the sum of six non-negative integers.

This is equivalent to the number of ways of representing 27 as the sum of non-negative integers which are all less than or equal to 6. (For an elementary proof of this see [10.1].)

This number is itself the coefficient of x^{27} in the expansion of

$$\prod_{n=1}^{n=6} (1 - x^n)^{-1}$$

and this can be found either (very laboriously!) by hand or by using a computer program. The number turns out to be 811.

If we insist (and it would be reasonable to do so) that every wheel must have at least one lug opposite it then the number is given by the coefficient of x^{21} in the expansion above because we can give each wheel one lug to start with and then distribute the remaining 21 without any further restrictions. This number turns out to be 331.

These are examples of a class of problems in the branch of mathematics known as combinatorics. Other examples include: given a positive integer N:

(1) In how many ways can N be represented as the sum of ***positive*** integers where the ***order of the integers is irrelevant?*** This number is denoted by $p(N)$ and called the number of *partitions* of N. For example:

$$4: = 4 = 3 + 1 = 2 + 2 = 2 + 1 + 1 = 1 + 1 + 1 + 1$$

and so, $p(4) = 5$.

There is no simple formula for the value of $p(N)$. For further details see [10.1].

(2) In how many ways can N be represented as the sum of (any number of) ***positive*** integers where the ***order of the integers is relevant?*** This number is denoted by $c(N)$ and is called the number of *combinations* of N. For example:

$$4{:}=4=3+1=1+3=2+2=2+1+1=1+2+1=1+1+2=1+1+1+1$$

so that $c(4)=8$. In fact it can be proved that

$$c(N)=2^{(N-1)},$$

for a proof of which see [10.2].

(3) In how many ways can N be represented as the sum of k (a *fixed number* of) *positive* integers when the *order of the integers is relevant?*
Since there are k integers and each of these must be greater than or equal to 1 the total number is given by the coefficient of X^N in the expansion of

$$X^k(1-X)^{-k}$$

or, what is the same thing, the coefficient of $X^{(N-k)}$ in the expansion of $(1-X)^{-k}$

and this is

$$\frac{(N-1)!}{(k-1)!(N-k)}$$

a formula which is relevant to Problem 4.2 (where $N=9$ and $k=3$, giving the value 28).

M19 Maximum multiple of the kick which can occur when differencing Hagelin key

Consider a wheel of length w with a kick of k. When we difference the pattern at any distance, d, which is not a multiple of w, there are four possibilities, shown in Table A.1.

Table A.1

Pin N	Pin $(N+d)$	Difference of key values
Inactive	Inactive	0
Inactive	Active	$+k$
Active	Inactive	$-k$
Active	Active	0

The extreme values of the differences are seen to be $\pm$k. If we now difference a second time, at any distance that is not a multiple of w, the extreme key differences that can occur are

$$k-(-k)=2k$$

and in the negative case

$$-k-(+k)=-2k.$$

Every subsequent differencing operation can at most double the previous extreme values. It follows that after n differencing operations the maximum multiples of the kick that can occur are

$$\pm 2^{(n-1)}k.$$

A somewhat similar situation arises in numerical computations where it takes the form:

> a single error in a table of values is propagated in an 'error fan'; the maximum error generated after n differencing operations will be

(the largest coefficient in the expansion of $(1+x)^n$) multiplied by (the original error).

So, for example, after six differencing operations a single error will have spread out and will be 20 times as large, in absolute value, in the centre of the sixth line of the error fan. For further information on such matters see [10.3].

M20 Determination of Hagelin slide by correlation coefficient

If the cage is known, the cryptanalyst can produce a 'theoretical cipher distribution' from the known frequencies of letters in the underlying plain language and the distribution of the frequencies of the 26 different key values, which implicitly assumes a slide of 0. The calculation is essentially the same as that described in M6. The actual frequencies of the letters in the cipher are then determined by counting. The two sets of frequencies ('theoretical cipher' and 'actual cipher') are now matched against each other at all 26 possible offsets and the correlation coefficient calculated in each case. In an ideal case the offset which yields the highest correlation coefficient should reveal the slide. In practice there may be more than one contender, but probably not many. Each would have to be tried. For details of the calculation of correlation coefficients see [2.4].

Chapter 13

M21 (Rate of increase of the number of primes)

The Prime Number Theorem [12.1] tells us that as N increases the number of primes less than N, which is traditionally denoted by $\Pi(N)$, is asymptotically approximated by

$$\Pi(N) \sim \frac{N}{\log(N)}$$

The logarithm being to base *e*.

It follows that as N *increases* the *fraction* of integers less than N which are primes slowly *decreases*. By studying tables of the number of primes less than 1000, 10 000, 100 000 Gauss discovered the Prime Number Theorem in 1793, but was unable to prove it. The relevant data are shown in Table A.2.

Table A.2

N	Number of primes less than N	Fraction of numbers which are prime
1000	168	1 in 5.95
10 000	1 229	1 in 8.14
100 000	9 592	1 in 10.43
1000 000	78 498	1 in 12.74

If we now difference the numbers in the right-hand column we get

$$8.14 - 5.95 = 2.19,$$
$$10.43 - 8.14 = 2.29,$$
$$12.74 - 10.43 = 2.31,$$

and Gauss conjectured that this difference would be essentially constant as N increased and would be approximately 2.3. Now log(10) is approximately equal to 2.3 and this implied that if we increase N by a factor of 10 the reciprocal of the fraction of integers less than N which are primes increases by log(10), a statement which is equivalent to the Prime Number Theorem. Gauss's conjecture was correct but it was more than a hundred years before the Prime Number Theorem was proved. See also [12.1].

M22 Calculating remainder using modular arithmetic

(1) That $(59)^{96}$ is a number with 171 digits follows from the fact that

$$96\log_{10}(59) = 96 \times (1.770\,85\ldots) = 170.0018\ldots$$

so $(59)^{96}$ lies between 10^{170} and 10^{171} and therefore has 171 digits.

(2) In using modular arithmetic it pays to remove the highest possible power of 2 from the exponent, then find the remainder of the other (odd) factor of the exponent, and finally use repeated squaring to get to the original exponent. So, for example, since $96 = 3 \times (32)$ if we form $(59)^3$ (mod 97) and repeatedly square it five times, reducing (mod 97) at each stage, we will obtain the required result; the details are as follows:

$$59 \times 59 = 3481 = 35 \times 97 + 86,$$

so

$$(59)^3 \equiv 86 \times 59 = 5074 = 52 \times 97 + 30 \equiv 30 \pmod{97},$$

so

$$(59)^6 \equiv (30)^2 = 900 = 9 \times 97 + 27 \equiv 27 \pmod{97},$$

so

$$(59)^{12} \equiv (27)^2 = 729 = 7 \times 97 + 50 \equiv 50 \pmod{97},$$

so

$$(59)^{24} \equiv (50)^2 = 2500 = 25 \times 97 + 75 \equiv 75 \pmod{97},$$

so

$$(59)^{48} \equiv (75)^2 = 5625 = 57 \times 97 + 96 = 96 \pmod{97} \equiv -1 \pmod{97},$$

and so, finally,

$$(59)^{96} \equiv (-1)^2 = 1 \pmod{97},$$

i.e. $(59)^{96}$ leaves remainder 1 when divided by 97, as asserted.

M23 Proof of the Fermat–Euler Theorem

It is helpful to begin with a proof of Fermat's Little Theorem; the generalistion to the Fermat–Euler Theorem is then almost obvious.

Fermat's Little Theorem asserts that

If p is a prime and M is any integer not divisible by p then $M^{(p-1)} \equiv 1 \pmod{p}$.

Proof

A complete set of residues ('remainders') (mod p) of numbers not divisible by p is

$$1, 2, 3, \ldots, (p-1).$$

Multiply each of these by M:

$$M, 2M, 3M, \ldots\ldots, (p-1)M.$$

No two of these numbers produce the same residue (mod p) for if, say,

$$aM \equiv bM \pmod{p}$$

then $M(a-b)$ is divisible by p; but M is not divisible by p and a and b are both less than p. Hence the $(p-1)$ multiples of M are all different (mod p); they must therefore be

$$1, 2, 3, \ldots, (p-1)$$

in some order. So

$$(M)(2M)(3M)\ldots((p-1)M) \equiv (1)(2)(3)\ldots(p-1) \pmod{p} = (p-1)! \pmod{p}.$$

Since $(p-1)!$ has no factor in common with p we can divide it out of both sides to give

$$M^{(p-1)} \equiv 1 \pmod{p}$$

which proves Fermat's Little Theorem.

Proof of the Fermat–Euler Theorem

We are now dealing with a composite modulus N. The proof follows along the same lines as above but now, instead of using *all* of the residues (mod p) we now consider only those residues which have no factor in common with N. If we denote these residues by

$$a_1, a_2, \ldots, a_k$$

then $k = \phi(\mathrm{N})$, where $\phi(\mathrm{N})$ is Euler's function, which was defined in M11. If we multiply each of the k residues by M then, as before, they are all different since if

$$M(a_r) \equiv M(a_s) \pmod{\mathrm{N}}$$

then $M(a_r - a_s)$ is divisible by N, but this is impossible since M has no factor in common with N and $(a_r - a_s)$ is less than N. We have therefore proved that

if M has no factors in common with N then
$M^{\Phi(\mathrm{N})} \equiv 1 \pmod{\mathrm{N}}$

which is the Fermat–Euler Theorem.

For the RSA encipherment system we need only the special case when $N=pq$, where p and q are different primes. In this case $\phi(N)=(p-1)(q-1)$.

M24 Finding numbers which are 'probably' primes

The 'sieve of Eratosthenes' will find *all* the primes below any given number N and, if a list of all the primes is what is required, this is the standard method. If, however, we only want to know *if a particular integer is a prime* then finding a list of all the primes below it is not necessary and, if the number is very large, likely to be very time-consuming. Unfortunately there is no very fast general method for testing if any given large number, N, is a prime and if N is large enough to be considered for the RSA method, say about 10^{50}, the time required to establish its primality beyond doubt is likely to be prohibitive. In view of this a different approach was proposed by Rabin in 1976 [12.8] and, in a different form, by Solovay and Strassen in 1977 [12.9]. The idea behind this is to use a test, which involves some number less than N, that

(1) will always fail if the number, N, *is* a prime,
(2) will succeed more often than not if N *is not* a prime.

When N is not a prime the test proposed by Rabin will succeed at least 75% of the time. If we use many numbers less than N, m say, and apply the test with each of these m and find that the test never succeeds then the probability that N is not prime is estimated to be $(0.25)^m$ and by taking m large enough the probability that N is prime can be made arbitrarily close to 1.

A description of Rabin's test can be found in [1.2], Chapter 9. For a bibliography of some relevant papers see [13.12].

M25 The Euclidean Algorithm

This algorithm is used to find the highest common factor (h.c.f.) of two integers, x_1 and x_2. If the h.c.f is denoted by h the algorithm can also be used to find integers m, n such that

$$mx_1 - nx_2 = h$$

which is relevant to the RSA encryption/decryption system.

The Euclidean Algorithm is carried out as follows.

We may suppose that both x_1 and x_2 are positive and that x_1 is greater than x_2; if not, interchange them.

Divide x_1 by x_2 to give a remainder x_3:

$$x_1 = a_1x_2 + x_3 \text{ where } a_1 \text{ is an integer and } 0 \leq x_3 < x_2.$$

If $x_3 \neq 0$ divide x_2 by x_3 to give a remainder x_4:

$$x_2 = a_2 x_3 + x_4 \text{ where } a_2 \text{ is an integer and } 0 \leq x_4 < x_3.$$

Continue in this way until the remainder term is 0, that is until we have

$$x_{(n-1)} = a_{(n-1)} x_n$$

Then h, the highest common factor of x_1 and x_2, is x_n. If $h = 1$ the integers x_1 and x_2 are said to be *relatively prime*.

Example
Find the h.c.f. of 1001 and 221.

Solution
$1001 = 4 \times 221 + 117,$
$221 = 1 \times 117 + 104,$
$117 = 1 \times 104 + 13,$
$104 = 8 \times 13 + 0.$

Therefore the h.c.f. of 1001 and 221 is 13. (Check: $1001 = 13 \times 91$; $221 = 13 \times 17$; and 91 and 17 are relatively prime.)

It is customary in mathematical literature to denote the h.c.f. of a pair of integers, m, n (say), by (m, n). Thus $(1001, 221) = 13$ and $(91, 17) = 1$.

The following example illustrates the use of the Euclidean Algorithm as needed in the RSA method of Chapter 13.

Example
Find integers m, n such that

$$91m - 17n = 1$$

Solution
We have

$91 = 5 \times 17 + 6,$
$17 = 2 \times 6 + 5,$
$6 = 1 \times 5 + 1,$
$5 = 5 \times 1,$

thus confirming that 91 and 17 are relatively prime (if they were not, the equation would have no solution). Working backward through the algorithm from the penultimate line:

$$1 = 6 - 1 \times 5 \text{ and } 5 = 17 - 2 \times 6$$

so

$$1 = 6 - 1 \times (17 - 2 \times 6) = 3 \times 6 - 17$$

but

$$6 = 91 - 5 \times 17$$

so

$$1 = 3 \times (91 - 5 \times 17) - 17 = 3 \times 91 - 16 \times 17.$$

Hence

$$m = 3 \text{ and } n = 16.$$

(Check: $3 \times 91 = 273$ and $16 \times 17 = 272$.)

An alternative method

The values of m and n can also be found by using *continued fractions* [6.7]. Although it looks different the method is essentially the same as that of the Euclidean Algorithm.

To illustrate the method we again find the values of m and n such that

$$91m - 17n = 1.$$

$$\frac{91}{17} = 5 + \frac{6}{17},$$

$$\frac{17}{6} = 2 + \frac{5}{6},$$

$$\frac{6}{5} = 1 + \frac{1}{5}.$$

The *partial quotients* of the continued fraction are therefore (5, 2, 1, 5) and its *convergents* are

$$\frac{5}{1}, \frac{11}{2}, \frac{16}{3}, \text{ and } \frac{91}{17}.$$

The numbers m and n are the numerator and denominator of the penultimate convergent: 16 and 3, as we found before.

M26 Efficiency of finding powers by repeated squaring

Given a number, X, that we wish to raise to the power n we could compute X^n by multiplying X by itself $(n-1)$ times. If n is small this is reasonable but if n is large it is very inefficient. Let k be such that

$$2^k < n < 2^{(k+1)};$$

then $k = [\log_2 n]$, where $[z]$ denotes, as usual in mathematics, the integer part of z.

If we compute $X^2, X^4, X^8, \ldots$ by repeated squaring we will need to carry out k squarings, that is k multiplications, to reach the power 2^k. The binary representation of n contains at most $(k+1)$ 1s and so X^n can be computed by multiplying together at most $(k+1)$ of the numbers $X, X^2, X^4, \ldots$ and this means that at most k further multiplications are required, giving a total of $2k$ multiplications in all.

Since $k < (\log_2 n + 1)$ we see that computing X^n by repeated squaring involves less than $2(\log_2 n + 1)$ multiplications whereas the brute force method requires $(n-1)$. If n is small the difference is not too great. When $n = 7$, for example, the brute force method requires 6 multiplications and the repeated squaring method requires 4. As n increases however the difference rapidly becomes very significant. When $n = 127$, for example, the brute force method requires 126 multiplications whereas repeated squaring needs only 12. For the really large exponents which are likely to occur in RSA encipherment/decipherment astronomical numbers of multiplications are replaced by a few hundred.

M27 Expected number of false hits in the 'meet-in-the-middle' attack on the DES

When we encipher a text using 2^{56} different keys we will obtain 2^{56} different encipherments. Since there are 2^{64} different 64-bit binary vectors there is only one vector in 256 $(=2^8)$ that will appear in the list of encipherments. The same is true when we *decipher* a text using 2^{56} different keys. If we now compare the two lists the chance that a vector in the encipherment list also occurs in the decipherment list is one in 256.

There are 2^{56} vectors in the encipherment list and one in 256 of them would be expected to appear in the decipherment list. We therefore expect 2^{48} agreements in all. All but one of these will be false, and one or more further tests must be applied to find the true solution.

M28 Elliptic Curve Cryptography

Despite the name the curves in question are not ellipses but are of the type

$$Y^2 = X^3 + aX + b$$

where a and b are integers. We are interested in pairs (X, Y) which are also integers; all arithmetic being carried out (mod p) for some (very large) prime p. Curves of this type can be parametrised by Weierstrass elliptic functions, hence the name.

So, for example, the points $(1, \pm 5)$ are integer points of the curve

$$Y^2 = X^3 + 2X + 3 \pmod{19}.$$

From any one or two points on the curve another may be constructed by using the tangent at the single point or the chord joining the two points. This tangent or chord meets the curve in a third point which must have rational co-ordinates and these rationals are convertible into integers in GF(p), the Galois field (mod p). So, for example, for the curve above with $p = 19$, the equation of the tangent at the point $(1, 5)$ is

$$2Y = X + 9$$

and we find that this tangent meets the curve again at the point where $X = -7/4$. This is equivalent to an integer value in GF(19); since 4 is the denominator of this fraction we must first find the integer n such that

$$4n \equiv 1 \pmod{19}.$$

This gives $n = 5$ since $20 = 1 \times 19 + 1$; hence $-7/4 \equiv (-7) \times 5 = -35 \equiv 3$ (mod 19) and so the fraction $-7/4$ is equivalent to the integer 3 in GF(19). This gives 3 as the integer value of X and the corresponding value of Y, obtained from the tangent above, is 6. Since

$$Y^2 = 36 \text{ and } X^3 + 2X + 3 = 27 + 6 + 3 = 36$$

we have verified that the points $(3, \pm 6)$ lie on the curve above. (We only need to show that they lie on the curve in GF(19); in fact they lie on the curve (mod p) for all p, but that is a fluke; this will not normally be the case.)

Thus another integer point is found on the curve. Since all arithmetic is (mod p) there are only a finite number of possible points (X, Y) with integer values. It follows therefore that the construction method that gives new points must eventually terminate. If we start with a particular (integer) point $Q(X, Y)$ on the curve we can generate a finite set, $\langle Q \rangle$, of points which we denote by $2Q, 3Q, 4Q, \ldots$ etc. (these are not to be confused with the points $(2X, 2Y)$ etc). For example, starting with the point $Q\,(1, 5)$ on the curve above we have just found the point, $2Q$, generated from the tangent at Q. Continuing in this way we find that we are led to the points

$$2Q = (3, \pm 6),\ 4Q = (10, \pm 4),\ 8Q = (12, \pm 8) \text{ and so on}$$

(for another example see [13.9]).

If we are given a point $R(X', Y')$ and asked to find if there is an integer n such that $R = nQ$, a point within the set $\langle Q \rangle$, we will have a very difficult problem unless the prime, p, is not too large. If R is not in the set $\langle Q \rangle$ no such value of n will be found. The values of p that are used are likely to exceed 10^{50} and the number of trials that will have to be made (except in some rare cases) is of the order of the square root of p, which makes the computational task beyond the power of even the most powerful computers.

The way in which Q, R and n are used to provide a signature to a message is somewhat involved and will not be described here. A reasonably brief and readable account will be found in [13.9].

Anyone wishing to know more about this particular aspect of Galois theory should consult books on finite fields. Galois was killed in a duel at the age of 20 in 1832. Knowing that he was almost certain to be killed he stayed awake during the night before the duel and wrote a paper, which he hoped would be published, explaining his ideas. The paper *was* eventually published, in 1846. Further details of his life and work will be found in books on the history of mathematics, such as [13.13].

Solutions to problems

Chapter 2

2.1 (Simple substitution)
The plaintext, with 'space' replacing the letter Z which was used before encipherment, is

```
A SOLEMN LITTLE REMINDER FROM AN ANCIENT POET
THE MOVING FINGER WRITES AND HAVING WRIT MOVES
ON NOR ALL THY PIETY NOR WIT SHALL LURE IT
BACK TO CANCEL HALF A LINE NOR ALL THY TEARS
WASH OUT A WORD OF IT
```

which is one verse from Edward Fitzgerald's translation of *The Rubáiyát of Omar Khayyám*.

Chapter 3

3.1 (Three Vigenère messages)
Since the plaintexts of the messages are identical, if we align the messages we can only get a cipher letter agreement if the letters of the keywords are also identical. The cryptanalyst ought to notice that cipher messages (1) and (2) have lots of agreements and that these all fall into three columns when the texts are written on a width of 8. The same happens to a lesser extent with cipher messages (2) and (3) when all the agreements fall in one column; but between cipher messages (1) and (3) there are no agreements at all. All this follows since the keywords are all 8 letters in length and

RHAPSODY and SYMPHONY agree in positions 4, 6 and 8;
SYMPHONY and SCHUBERT agree only in position 1;
RHAPSODY and SCHUBERT have no identical letters in the same position.

The encipherments of the message using the three keywords when written in three lines under each other in blocks of 8 letters are as follows:

```
EVWMAGAR YLXIAAHV WVRMSZOV XVOSPAHL
FMIMPGKR ZCJIPARV XMDMHZYV YMASEARL
FQDRJWOM ZGENJQVQ XQYRBPCQ YQVXYQVG

OAOMUCPC OAOMLVHV RPDMGTAR YLXESFWW
PRAMJCZC PRAMAVRV SGPMVTKR ZCJEHFGW
PVVRDSDX PVVRULVQ SKKRPJOM ZGEJBVKR
```

Messages (1) and (2), and (2) and (3) are in 'partial depth'.
Such an observation ought to quickly lead to a solution.

3.2 (Vigenère decryption)
Studying the cipher text reveals several digraphs which occur four times or more and among these are some which extend to repeats of three or four letters, including: ZMUI which occurs at positions 15 and 135; ZMUE at positions 67 and 163; and KRD at positions 9, 8, 172 and 176. All the intervals between these repeats are multiples of 4 so we conclude that the key is of length 4.

Looking at the four cipher letter frequency distributions we find that the cipher letter for 'space' is almost certainly M in the first alphabet, Z in the third and S in the fourth; the second alphabet is slightly less informative but D is the best bet and so we are led to the key as probably being 13-4-0-19 which is equivalent to the keyword NEAT.

Decryption of a few words confirms this and, with 'space' replacing Z, the decrypt is

THERE ARE SOME THEOREMS WITH A PROOF WHICH IS SO SHORT AND ELEGANT THAT IT SEEMS UNLIKELY THAT A BETTER ONE WILL EVER BE FOUND SUCH IS THE CASE WITH EUCLIDS PROOF THAT THERE ARE AN INFINITE NUMBER OF PRIMES THE PROOF IS IN THE APPENDIX IN THIS BOOK. (M4)

Chapter 4

4.1 (Simple transposition)

If the key length is 6 each column of the transposition box will contain five letters; we therefore write the cipher text out in columnar form in Table S.1.

Table S.1

1	2	3	4	5	6
L	S	L	A	H	I
P	C	A	M	O	R
E	E	E	H	T	T
U	O	M	S	A	M
D	E	A	S	R	Y

The third line looks as if it might contain the trigraph THE which suggests that either column 5 or column 6 should be immediately to the left of column 4. Examination of the other digraphs in 5-4 and 6-4 points to 5-4 as being more likely. If the ordering 5-4 *is* correct and the word THE is present then column 4 must precede column 1, 2 or 3 which doesn't help us much at this stage. We therefore look elsewhere and try to find which column might precede column 5 we see that the only digraph that looks plausible in row 1 is SH which implies that column 2 should be immediately to the left of column 5. We therefore have a partial tentative ordering

2-5-4

and if we write out those three columns in that order we have Table S.2.

Table S.2

2	5	4
S	H	A
C	O	M
E	T	H
O	A	S
E	R	S

The solution now follows fairly easily. The key is 2 5 4 1 3 6 and the plaintext, with spaces inserted, is

SHALL I COMPARE THEE TO A SUMMERS DAY.

4.2 (Number of possible transposition boxes)
In the particular case of nine letters in three columns the possibilities are:

7, 1, 1; 1, 7, 1; 1, 1, 7;
6, 2, 1; 6, 1, 2; 2, 6, 1; 2, 1, 6; 1, 6, 2; 1, 2, 6;
5, 3, 1; 5, 1, 3; 3, 5, 1; 3, 1, 5; 1, 5, 3; 1, 3, 5;
5, 2, 2; 2, 5, 2; 2, 2, 5;
4, 4, 1; 4, 1, 4; 1, 4, 4;
4, 3, 2; 4, 2, 3; 3, 4, 2; 3, 2, 4; 2, 4, 3; 2, 3, 4;
3, 3, 3.

A total of 28 (no column of the box is allowed to have 0 letters).
This is a particular case of a more general problem:

> In how many ways can n be represented as the sum of k positive integers when the order of the integers is relevant?

It can be shown (see M18) that the number is

$$\frac{(n-1)!}{(k-1)!(n-k)!}$$

Putting $n=9$ and $k=3$ gives us

$$\frac{8!}{2!6!}=\frac{8\times 7}{2}=28.$$

When $n=35$ and $k=5$ the corresponding figure is $(34\times 33\times 32\times 31)/24 = 46\,376$.

4.3 (Boustrophedon rows in a transposition box)
Alternate vertical digraphs at the ends of the rows will be unaltered in the cipher text.

Chapter 5

5.1 (MDTM)
The cipher text is

```
CFIGS FLTBC XKEEA EBHTB GLDPI
```

and the 5×5 substitution box is shown in Table S.3.

Table S.3

	A	B	C	D	E
A	A	B	S	O	L
B	U	T	E	C	D
C	F	G	H	I	K
D	M	N	P	Q	R
E	V	W	X	Y	Z

We begin the decryption by converting the monographs back into digraphs:

BDCAC DCBAC CAAEB BABBD ECCEB CBCAA BCABC
CBBAB CBAEB EDCCD.

Table S.4

3	1	5	2	4
E	B	C	C	B
C	D	B	A	C
C	C	A	A	A
E	A	E	E	B
B	C	B	B	C
C	D	E	B	C
B	C	D	A	B
C	B	C	B	B
A	A	C	B	A
A	C	D	D	B

The transposition is 3-1-5-2-4 so we write this text vertically into a rectangle with five columns in the column order given by the transposition: see Table S.4. Finally we recover the plaintext by reading the text row by row and converting the digraphs back to monographs using the 5×5 square which produces the text

WHENSHALLWETHREEMEETAGAIN

or, inserting spaces,

WHEN SHALL WE THREE MEET AGAIN

– the opening line of Shakespeare's *Macbeth*.

5.2 (Playfair)

With the keyword RHAPSODY the Playfair encipherment square is as shown in Table S.5

Table S.5

R	H	A	P	S
O	D	Y	B	C
E	F	G	I	K
L	M	N	Q	T
U	V	W	X	Z

and, breaking the cipher text into digraphs, we have

OX BG IH PE OK GH MT TR OI UE VG KG NC

We convert these to plaintext using the Playfair square, which produces

BU YI FP RI CE FA LQ LS BE LO WF IF TY

or, after running the text together, inserting spaces between words and deleting the single dummy Q,

BUY IF PRICE FALLS BELOW FIFTY.

Chapter 6

6.1 (Fibonacci key)

(1) Starting with 0 and 2 as the first two terms produces a sequence which repeats after 20 digits:

0, 2, 2, 4, 6, 0, 6, 6, 2, 8, 0, 8, 8, 6, 4, 0, 4, 4, 8, 2, 0, 2, 2,...

All the terms are *even* and the sequence is quite unsuitable as a key.

(2) Starting with 1 and 3 produces a sequence which repeats after 12 digits:

1, 3, 4, 7, 1, 8, 9, 7, 6, 3, 9, 2, 1, 3, 4,

6.2 (Code plus additive)

Cipher text	86	69	42	19	60	35	08	13	76	48	23	02	50	91
Key	12	31	35	45	84	94	37	37	18	07	98	74	86	15
Difference	74	38	17	74	86	41	71	86	68	41	35	38	74	86
Text	T	H	A	T	X	I	S	X	R	I	G	H	T	X

i.e. THAT IS RIGHT with X separating the words.

Chapter 7

7.1 (Stencil cipher solutions)
We make a frequency count of the letters of the text and of the letters of the four 'possible solutions'. If the letters of a 'possible solution' can all be found with a frequency no higher than that of the corresponding letter of the text then the 'possible solution' is indeed possible, otherwise it isn't. The five frequency counts are shown in Table S.6 and these show that the third of these is not a possible solution since it has a letter, W, which does not occur at all in the text. The other three *are* possible solutions.

Table S.6

	A	B	C	D	E	F	G	H	I	J	K	L	M	N	O	P	Q	R	S	T	U	V	W	X	Y	Z
Text	16	1	4	3	15	3	1	7	3	0	2	7	4	8	6	5	1	10	10	8	3	0	0	0	3	0
(1)	1	0	1	0	1	0	0	1	1	0	0	0	2	0	0	0	0	3	2	1	0	0	0	0	1	0
(2)	1	0	2	0	2	0	0	0	0	0	0	0	1	1	2	0	0	0	0	1	0	0	0	0	0	0
(3)	2	0	1	0	0	0	1	0	1	0	1	1	0	0	1	0	1	0	0	0	1	0	1	0	2	0
(4)	2	1	1	0	2	1	0	3	1	0	0	2	0	1	5	0	0	1	1	3	1	0	0	0	0	0

7.2 (Decrypt of a book cipher)
Using Table 7.3 we obtain the decrypt:

THEXSUSPECTXHASXMOVEDXTOXLIVERPOOLX.

7.3 (Continuation of example solution)
Continuing the decrypt which was started in the example we get

Key	SXAREXSPRINGXFLOWERSX
Message	MOREXFUNDSXURGENTLYXX

Since the key evidently refers to daffodils and the erroneous text went wrong at the fourth letter we guess that the sender left out the second F of the keyword DAFFODILS and the full key and message, with spaces restored in place of X, are

Key	DAFFODILS ARE SPRING FLOWERS
Message	WE NEED MORE FUNDS URGENTLY

Chapter 8

8.1 (Recurrences of order 4)
(i) Taking $U_0 = U_1 = U_2 = U_3 = 1$ the recurrence

$$U_n = U_{(n-1)} + U_{(n-4)}$$

produces the sequence

1, 1, 1, 1, 0, 1, 0, 1, 1, 0, 0, 1, 0, 0, 0, 1, 1, 1, 1, ...

which repeats after 15 terms but not before.

(ii) With the same starting values the recurrence

$$U_n = U_{(n-1)} + U_{(n-2)} + U_{(n-3)} + U_{(n-4)}$$

produces the sequence

1, 1, 1, 1, 0, 1, 1, 1, 1, ...

which repeats after only 5 terms.

8.2 (Cycling in a mid-squares random number generator)
Starting with $X = 7789$ the sequence continues

6685, 6892, 4996, 9600, 1600, 5600, 3600, 9600, ...

a cycle of length 4.

8.3 (Cycle lengths in linear congruences)
(1) The congruence $U_n = 3U_{(n-1)} + 7 \pmod{19}$ beginning with $U_0 = 1$ continues

10, 18, 4, 19, 7, 9, 15, 14, 11, 2, 13, 8, 12, 5, 3, 16, 17, 1, ...

The cycle is of length 18 and is maximal since the value 6 cannot occur, because it produces a cycle of length 1.

(2) The congruence $U_n = 4U_{(n-1)} + 7 \pmod{19}$ beginning with $U_0 = 1$ continues

11, 13, 2, 15, 10, 9, 5, 8, 1, ...

The cycle length is 9 and is not maximal. No recurrence with multiplier 4 is maximal when the modulus is 19.

Chapter 9

9.1 (Mini-Enigma)
The enciphered doublets yield the chains

0239, 1648, 55 and 77.

The encipherment table for R1, given column 1 and using the diagonal property, is shown in Table S.7.

Table S.7

	Setting									
	1	2	3	4	5	6	7	8	9	10
0	0	3	1	8	5	2	9	1	4	7
1	8	1	4	2	9	6	3	0	2	5
2	6	9	2	5	3	0	7	4	1	3
3	4	7	0	3	6	4	1	8	5	2
4	3	5	8	1	4	7	5	2	9	6
5	7	4	6	9	2	5	8	6	3	0
6	1	8	5	7	0	3	6	9	7	4
7	5	2	9	6	8	1	4	7	0	8
8	9	6	3	0	7	9	2	5	8	1
9	2	0	7	4	1	8	0	3	6	9

The 1-chains tell us immediately that (5, 7) is a pair. The two 4-chains can be aligned (with one reversed) in eight ways. Normally we would have to try each of these but we shall only align them in the correct way, viz:

1 8 4 6
9 0 2 3

Table S.8

Pair	Setting 1	Pair	Setting 2
(5, 7)	(7, 5)	(5, 7)	(4, 2)
(1, 9)	(8, 2)	(1, 0)	(1, 3)
(8, 0)	(9, 0)	(8, 2)	(6, 9)
(4, 2)	(3, 6)	(4, 3)	(5, 7)
(6, 3)	(1, 4)	(6, 9)	(8, 0)

If we now encipher the vertical and diagonal pairs at settings (1 and 2) they give Table S.8, and the two sets of five pairs, listed in the right-hand columns in each case, are full of contradictions; so we reject the hypothesis that R1 was set at position 1 at the encipherment of the indicators. If we now encipher the same pairs at settings (3 and 4) we get Table S.9, and these pairings are completely consistent. Any other combination of chain alignment and setting of R1 would lead to contradiction. We therefore conclude that R1 was set at position 3 at the start of the encipherment of the indicators at the ground setting. If this is so, the pairings in the composite reflector are (0, 5), (1, 3), (2, 8), (4, 7), and (6, 9).

Table S.9

Pair	Setting 3	Pair	Setting 4
(5, 7)	(6, 9)	(5, 7)	(9, 6)
(1, 9)	(4, 7)	(1, 0)	(2, 8)
(8, 0)	(3, 1)	(8, 2)	(0, 5)
(4, 2)	(8, 2)	(4, 3)	(1, 3)
(6, 3)	(5, 0)	(6, 9)	(7, 4)

Chapter 10

10.1 (Hagelin message)

Since the cage is (0, 5, 5, 5, 5, 5) the only keys that can occur are 0, 5, 10, 15, 20 and 25 with relative frequencies 1, 5, 10, 10, 5 and 1. We write down six lines showing what the 'plaintext' would be if the key value were 0, 5, 10, 15, 20 or 25 at each position of the text using the Hagelin decipherment rule:

plaintext letter = key − cipher letter.

The cipher and six lines corresponding to the six keys are shown in Table S.10.

Table S.10

Cipher	CBZPC CJXWY CXSHN IQUSR
Key = 0	YZBLY YRDEC YDITN SKGIJ
Key = 5	DEGQD DWIJH DINYS ^PLNO
Key = 10	IJLVI IBNOM INSD^ CUQST
Key = 15	NOQAN NGSTR NS^IC HZV^Y
Key = 20	STVFS SL^YW S^CNH MEACD
Key = 25	^YAK^ ^UCDB ^CHSM RJFHI

The solution is

```
SOLVING THIS IS EASY
```

The numbers of correct plaintext letters appearing in the six rows are seen to be

0, 3, 5, 8, 4, and 0

These agree fairly well with what we would expect from the theoretical key distribution which predicts that, in a sufficiently long text, the ratios should be

1:5:10:10:5:1.

10.2 (Hagelin cages)

All but cages (b) and (e) generate all key values (mod 26). Cage (b) fails to generate key values 13 and 14. Cage (e) fails to generate key values 4, 12, 15 and 23. Note that if an unoverlapped cage which uses 27 lugs fails to generate key value N it must also fail to generate key value $(27 - N)$ since the kicks add up to 27 and by reversing the pins on the wheels a key value of N becomes a key value of $(27 - N)$.

10.3 (Overlapped Hagelin cage producing key value 17)

There is unfortunately no short cut to finding these representations although we can reduce the number of pin combinations that we need to examine by noting that one, and only one, of the two 'big' kicks (11, 9) will be required, since together they give 18 (ie $11 + 9 - 2$), and the other four (7, 5, 3 and 1) only add up to 16, and overlaps would reduce this further. We need therefore consider only 32 of the 64 possible pin combinations.

The six combinations which produce key value 17 are:

OXXOXO	which gives	$(9+7+3)-(2)=17;$
OXXOXX	which gives	$(9+7+3+1)-(2+1)=17;$
OXXXOO	which gives	$(9+7+5)-(2+2)=17;$
XOOXOX	which gives	$(11+5+1)-(0)=17;$
XOOXXO	which gives	$(11+5+3)-(2)=17;$
XOOXXX	which gives	$(11+5+3+1)-(2+1)=17.$

Chapter 11

11.1 (Pin-setting errors in the Hagelin and SZ42)
(1) Every 23rd cipher letter would be different in the two cipher texts. In an unoverlapped machine the amount of the difference would be the kick on the 23-wheel. In an overlapped machine the difference might be one of two or more values, depending upon how the 23-wheel was overlapped with the other wheels.

(2) Every 31st cipher letter would be different in the two cipher texts, the difference being seen in the 2nd bit of the 5-bit ITA characters.

(3) Since the 61-wheel controls the 37-wheel, which in turn controls all the wheels of set C, the 37-wheel, and hence the wheels of set C, would get steadily more and more out of step with their correct positions and so the cipher texts would differ, apart possibly from an occasional accidental agreement, from some point in the first 61 letters of the text.

Chapter 13

13.1 (Self-encipherment in the RSA system)
We have to compute 530^{17} and 531^{17} (mod 3127). This is easy since $17=16+1$. We therefore find the 16th power of each by squaring four times.

$$530^2=280\,900=89\times3127+2597\equiv2597 \pmod{3127}$$

so

$$530^4\equiv2597^2=6\,744\,409=2156\times3127+2597\equiv2597 \pmod{3127};$$

it therefore follows that $530^{16}\equiv2597^8\equiv2597 \pmod{3127}$ and hence

$$530^{17}\equiv530\times2597=1\,376\,410=440\times3127+530\equiv530 \pmod{3127},$$

that is, 530 enciphers to itself in this RSA system.

In the case of 531 we have

$$531^2=281\,961=90\times3127+531\equiv531 \pmod{3127}$$

and therefore

$$531^{17}=531\times(531)^{16}\equiv531\times531\equiv531 \pmod{3127},$$

that is, 531 also enciphers to itself in this RSA system.

References

Chapter 1

[1.1] Hill, R.: *A First Course in Coding Theory*, Oxford University Press (1986). Assumes high school mathematics, including matrices.

[1.2] Welsh, Dominic: *Codes and Cryptography*, Oxford Science Publications (1988). Undergraduate level, mathematics, computer science etc.

[1.3] Berlekamp, E. R.: *Algebraic Coding Theory*, McGraw-Hill, New York (1968). Pretzel, Oliver: *Error-Correcting Codes and Finite Fields*, Clarendon Press, Oxford (1992). Postgraduate-level texts.

[1.4] Chadwick, J.: *The Decipherment of Linear B*, Cambridge University Press (1958). Describes the solution, in 1952, by Michael Ventris (who was killed in an accident in 1956). A celebrated earlier feat was the decipherment of the hieroglyphics of the Rosetta Stone by J. F. Champollion, but he, unlike Ventris, had the advantage of a 'parallel text' in a known language. Anyone interested in attempting to decipher unsolved scripts might care to investigate the Beale Ciphers or Voynich Manuscript; but such endeavours are not for the faint-hearted. At a more serious level there are texts in various unknown languages, including Linear A, awaiting solution.

[1.5] Public Records Office: *GARBO: The Spy Who Saved D-Day*, PRO, Richmond, Surrey (2000). Appendix XXXVI deals with 'Secret inks'.

Chapter 2

[2.1] In 1968 the author was asked to decipher part of the diary that Wittgenstein kept during his time in the Austrian Army in the 1914–18 war. It was hoped that it might provide clues as to some of his philosophical ideas, but the pages deciphered contained only comments on the misery of army life. If the diary has been published I am not aware of it.

[2.2] Francis, W. N.: *A Standard Sample of Present-Day Edited American English*, Brown University, Providence, Rhode Island (1964).

[2.3] Wright, E.V.: *Gadsby: A Story of Over 50,000 Words without Using the Letter E*, Weteel, Los Angeles (1939).

[2.4] Moroney, M.J.: *Facts from Figures*, Pelican Books, Harmondsworth, Middlesex (1951). This deservedly popular book was reprinted many times but is now out of print. If a copy can be found it is worth consulting. It introduces the reader to the main tools of statistical analysis and is suitable for readers who do not have an advanced knowledge of mathematics. Correlation coefficients are covered in Chapter 16. If unavailable see almost any other book on elementary statistics.

[2.5] Hinsley, F.H. and Alan Stripp (eds): *Codebreakers. The Inside Story of Bletchley Park*, Oxford University Press (1993). See Chapter 19 (by I.J. Good). The formula is quoted on page 150.

[2.6] Hall, Marshall, Jr: *Combinatorial Theory*, John Wiley & Sons, New York (1986).

Chapter 3

[3.1] Hinsley, F.H. and Alan Stripp (eds): [2.5]. Jack Good's account of both the 8-letter bogus repeat and the 22-letter genuine repeat are in Chapter 19, pages 156 and 157.

Chapter 4

[4.1] Hinsley, F.H. and Alan Stripp (eds): [2.5]. This is also reported by Jack Good in Chapter 19, page 160.

Chapter 5

[5.1] Clark, R.W.: *The Man Who Broke Purple*, Weidenfeld & Nicolson, London (1977). This is an account of the career of William F. Friedman, the greatest American cryptanalyst of his day. The ADFGVX and other ciphers are described on pages 43–6.

[5.2] Hinsley, F.H. and Alan Stripp (eds): [2.5]. The JN40 and other Japanese naval codes and ciphers are described in the article by Hugh Denham (Chapter 27).

[5.3] Konheim, A.G.: *Cryptography. A Primer*, John Wiley & Sons, New York (1981).

[5.4] Hinsley, F.H. and Alan Stripp (eds): [2.5]. The German Double Playfair system is covered in Chapters 22 and 23.

Chapter 6

[6.1] Tuchman, Barbara: *The Zimmermann Telegram*, Ballantine, New York (1994).

[6.2] Hinsley, F.H. and Alan Stripp (eds): [2.5]. See Chapter 6.

[6.3] Hinsley, F.H. and Alan Stripp (eds): [2.5]. Japanese naval codes and ciphers are described in Chapter 27.

[6.4] Kahn, David: *Seizing the Enigma*, Souvenir Press, London (1991), pages 203–5.

[6.5] Vajda, S.: *Fibonacci and Lucas Numbers and the Golden Section. Theory and Applications*, Ellis Horwood, Chichester (1989).

[6.6] *The Fibonacci Quarterly* is published by the Fibonacci Association.

[6.7] Almost any elementary book on the theory of numbers will include an account of continued fractions, which are used for finding good rational approximations to irrational numbers. Hardy and Wright's book, [10.1], is particularly good.

Chapter 7

[7.1] Public Records Office: [1.5]. Appendix XXXIII covers GARBO's ciphers and transmitting plans.

Chapter 8

[8.1] Feller, William: *An Introduction to Probability Theory and Its Applications*, Volume 1, 3rd Edition, John Wiley & Sons, New York (1972).

[8.2] Golomb, S.W.: *Shift Register Sequences*, Holden-Day, San Francisco (1967).

[8.3] Hammersley, J.W. and D.C. Handscomb: *Monte Carlo Methods*, Methuen, London (1964).

[8.4] Press, W.H., B.P. Flannery, S.A. Teukolsky and W.T. Vetterling: *Numerical Recipes*, Cambridge University Press (1986).

Chapter 9

[9.1] Churchhouse, R.F.: 'A classical cipher machine: the ENIGMA – some aspects of its history and solution', *Bulletin of the Institute of Mathematics and Its Applications*, 27 (1991), 129–37. (A proof of the 'chaining theorem' is given on page 134.)

[9.2] Hinsley, F.H. and Alan Stripp (eds): [2.5]. Chapters 11–17 (pages 81–137) are devoted to the Enigma.

[9.3] Garlinski, J.: *Intercept. The ENIGMA War*, Magnum Books (Methuen), London (1981). In addition to covering the history of the Enigma and its solution this book contains an appendix by Colonel Tadeusz Lisicki which explains the Polish method of solution in some detail.

[9.4] Deavours, C.A.: *Breakthrough '32: The Polish Solution of the ENIGMA*, Aegean Park Press, Laguna Hills, California (1988).This booklet includes an MS DOS diskette for an IBM PC (in BASIC and machine language) with fully worked examples, wheel wirings etc.

[9.5] Hinsley, F.H. and Alan Stripp (eds): [2.5]. Chapter 16 (pages 123–31) by Peter Twinn deals with the Abwehr Enigma.

Chapter 10

[10.1] Hardy, G.H. and E.M. Wright: *An Introduction to the Theory of Numbers*, Oxford University Press. The chapter on 'Partitions' in any of the editions. The identity referred to is a classical elementary theorem in combinatorics.

[10.2] Andrews, G.E.: *The Theory of Partitions*, Addison-Wesley, Reading, Massachusetts (1976). The formula for the number of compositions of a number is given in Example 3 on page 63.

[10.3] Almost any elementary book on numerical methods or numerical analysis will include discussion of error propagation. A table showing the error fan illustrating the point referred to in M19 is given on page 63 in an early book on the use of computers in numerical work: National Physical Laboratory: '*Modern Computing Methods*', Notes on Applied Science No. 16, HMSO, London (1961).

[10.4] Barker, W.G.: *Cryptanalysis of the Hagelin Cryptograph*, Aegean Park Press, Laguna Hills, California (1977).

[10.5] Beker, H. and F. Piper: *Cipher Systems*, Northwood Books, London (1982).

Chapter 11

[11.1] Hinsley, F.H. and Alan Stripp (eds): [2.5]. The ITA is given in Chapter 19.

[11.2] Beker, H. and F. Piper: [10.5]. See Appendix 2.

[11.3] Hinsley, F.H. and Alan Stripp (eds): [2.5]. Chapters 18–21 deal with the SZ42 and Colossus.

Chapter 12

[12.1] For a *statement*, but not a *proof*, of the Prime Number Theorem see Hardy and Wright, [10.1]. Chapter 1. There is no simple proof of this theorem. It was conjectured by Gauss in 1793 but wasn't proved until 1896, by Hadamard and de la Vallée Poussin independently. The usual proof involves complex variable theory and will be found in books on analytic number theory such as Ingham, A.E.: *The Distribution of Prime Numbers*, Cambridge Tract Number 30 (Reprinted, Hafner, 1971) or Estermann, T.: *Introduction to Modern Prime Number Theory*, Cambridge Tract Number 41 (1952).

[12.2] Metropolis, N., J. Howlett and G-C. Rota (eds): *A History of Computing in the Twentieth Century*, Academic Press, New York (1980). This excellent collection of essays, written by more than 30 authors, covers the machines, the programming languages and the people in various countries including the USA, UK, Germany, USSR and Japan.

[12.3] *IBM Journal of Research and Development 25th Anniversary Issue*, 25, 5 (September 1981). This is a comprehensive account of the IBM computers, and the relevant technology, produced from the 1950s to 1980.

[12.4] Lavington, S.H.; *A History of Manchester Computers*, NCC Publications, Manchester (1975). An account of the computers produced collaboratively by Manchester University, Ferranti Ltd and ICL between 1946 and 1974.

[12.5] Churchhouse, R.F.; 'Experience with some early computers', *Computing & Control Engineering Journal* (April 1993), 63–7. An account of the computers programmed by the author (Manchester Mark 1, IBM 704, Univac 1103A) in the 1950s.

[12.6] Diffie, W. and M.E. Hellman: 'New directions in cryptography', *Transactions of the IEEE on Information Theory*, **IT-22**, No. 6, 644–54 (November 1976), but see also [13.1].

[12.7] Davies, D.W. and W.L. Price: *Security for Computer Networks*, John Wiley & Sons, Chichester (1984). Chapter 8 deals with various aspects of public key cryptography.

[12.8] Rabin, M.O.: 'Probabilistic algorithms' in Traub, J.F. (ed): *Algorithms and Complexity*, Academic Press, New York, pages 21–39 (1976).

[12.9] Solovay, R. and V.Strassen: 'A fast Monte Carlo test for primality', *SIAM Journal of Computing*, **6**, (1977), 84–5. Erratum: *ibid*, **7** (1978), 118.

Chapter 13

[13.1] Rivest, R.L., A.Shamir and L. Adelman: 'A method for obtaining digital signatures and public key cryptosystems', *Communications of the ACM*, **21** (1978), 120-26. The essentials of this method and the Diffie–Hellman key exchange system had been discovered some years earlier by James Ellis at GCHQ but security restrictions prevented publication at the time. For an account of the work of Ellis and others at GCHQ see the article 'The open secret" by Steven Levy in *Wired*, April 1999, 108–15.

[13.2] The proof of Fermat's Last Theorem, by Andrew Wiles in 1993, involves extremely sophisticated mathematics. For a good, very readable, account see Singh, Simon: *Fermat's Last Theorem*; Fourth Estate, London (1997).

[13.3] National Bureau of Standards in *Federal Register*, issue of May 15th, 1973.

[13.4] National Bureau of Standards in *Federal Register*, issue of August 27th, 1974.

[13.5] Konheim, A.G.: [5.2]. A full description of the DES with results of test data is given on pages 240–79.

[13.6] Davies, D.W and W.L. Price; [12.7] Chapter 3, pages 49–87 deals with the DES. It should be noted that the S-Box tables on page 58 contain three errors. The S-Boxes are given correctly in [13.7].

[13.7] Federal Information Processing Standards Publication **185**: Escrowed Encryption Standard (EES). A complete specification of Skipjack is given in 'Skipjack and KEA Algorithm Specifications' (Version 2.0, May 1998).

[13.8] Lai, X. and J.L. Massey: 'A proposal for a new block encryption standard', *Advances in Cryptology, Eurocrypt '90*, Springer-Verlag, Berlin, pages 389–404; (1991).

[13.9] Galbraith, S.: 'Elliptic curve public key cryptography', *Mathematics Today*, **35** (3), 76–9 (June 1999).

[13.10] Zimmermann, Philip: *The Official PGP User's Guide*, MIT Press (1996).

[13.11] Garfinkel, Simson L: *PGP. Pretty Good Privacy*, O'Reilly, Sebastopol, California (1994).

[13.12] Beauchemin, P., G. Brassard, C. Crepeau, C. Goutier, and C. Pomerance: 'The generation of random numbers that are probably prime', *Jounal of Cryptology*, **1**, 53–64 (1988).

[13.13] Bell, E.T; *Men of Mathematics*, Pelican Books, Harmondsworth, Middlesex (1953). Originally published in 1937. Provides very readable accounts of the lives and works of more than 30 of the greatest mathematicians from ancient times to the early twentieth century. This edition is in two volumes; the chapter on Galois is in Volume 2.

Name index

Adelman, L. 171, 234
Andrews, G. E. 232

Barker, W. G. 233
Beauchemin, P. 234
Beker, H. 233
Bell, E. T. 234
Berlekamp, E. R. 230
Brassard, G. 234
Brooke, R. 78

Caesar, Julius *passim*
Chadwick, J. 230
Champollion, J. F. 230
Churchhouse, R. F. 232, 233
Clark, R. W. 231
Crepeau, C. 234

Davies, D. W. 233, 234
Deavours, C. A. 232
Denham, H. C. 231
Dickens, C. 33
Diffie, W. 166, 233
Doyle, A. C. ix

Ellis, J. 234
Estermann, T. 233
Eratosthenes 173
Euclid 192–3
Euler, L. 173, 175, 205, 211

Feller, W. 232
Fermat, P. 173, 175, 234
Fibonacci *passim*
Fitzgerald, E. 218
Flannery, B. P. 232
Francis, W. N. 230
Friedman, W. F. 231

Galbraith, S. 234
Galois, E. 216–7
'GARBO' 9, 52, 88–92, 230, 232
Garfinkel, S. L. 234
Garlinski, J. 232
Gauss, C. F. 209, 233
Golomb, S. W. 232
Good, I. J. 31, 35, 132, 231
Goutier, C. 234

Hadamard, J. 233
Hammersley, J. W. 232
Hamming, R. W. 8
Handscomb, D. C. 232
Hardy, G. H. 232, 233
Hellman, M. E. 166, 233
Hill, R. 230
Hinsley, F. H. 231 and *passim*
Howlett, J. 233

Ingham, A. E. 233

Jefferson, T. 37–9, 110, 122

Kahn, D. 231
Konheim, A. G. 231, 234

Lai, X. 187, 234
Lavington, S. 233
Leonardo of Pisa *see* Fibonacci
Levy, S. 234

Massey, J. L. 187, 234
Metropolis, N. 233
Moroney, M. J. 231
Morse, S. 64

Painvin, G. 58
Pepys, S. 7

Piper, F. 233
Poe, E. A. ix
Pomerance, C. 234
Press, W. H. 232
Price, W. L. 233, 234
Pujol, J. *see* GARBO

Rabin, M. O. 212, 234
Rivest, R. L. 171, 234
Rota, G.-C. 233

Scherbius, A. 111
Shamir, A. 171, 234
Singh, S. 234
Solovay, R. 212, 234
Strassen, V. 212, 234
Stripp, A. 231 and *passim*

Teukolsky, S. A. 232
Tuchman, B. 231
Turing, A. M. 161

Vajda, S. 231
Vallée-Poussin, C. de la 233
Ventris, M. 7, 230
Vetterling, W. T. 232

Welsh, D. 230
Wiles, A. 234
Wittgenstein, L. 23, 230
Wright, E. M. 232, 233
Wright, E. V. 230

Zimmermann, A. 65, 231
Zimmermann, P. 187

Subject index

Abwehr Enigma 124, 132
active pins 136
algorithm 5
amplifier noise 97–8
anagram ix, 40
arithmetic, modular 10, 68, 209
authentication 165, 188–9
avalanche test (DES) 184

Beale cipher 230
binary key stream 100–3
  improving security 104–6
binary linear recurrence 198–202
birthdays paradox 191–2
Bletchley ix, 131, Plate 11. 1
book cipher 36, 75–87
  decipher table 78
  disastrous error in using 86
  encipher table 77
  letter frequencies in 79, 194–5
  solving 79–85
breaking ciphers 4
Brown corpus 23, 42
brute force attack 27, 44, 142, 158, 187, 215

C36 cipher machine 133
C38 cipher machine 133
C41 cipher machine 133
Caesar's cipher Chapter 2 and *passim*
cage
  'good' 141
  number of 141, 206–7
cathode ray tube storage 161
cipher
  1918 German Army 57
  Beale 230
  block 183
  book *see* book cipher
  Caesar's *see* Caesar's cipher
  digraph-to-digraph 58
  GARBO's *see* GARBO
  German double Playfair 61–2
  Japanese naval 58
  Jefferson's cylinder 37–9
  jigsaw *see* jigsaw cipher
  MDTM 56
  monograph-to-digraph 54
  Playfair 59–60
  simple substitution Chapter 2 and *passim*
  stencil 73–4
  transposition Chapter 4
  two-letter Chapter 5
  unbreakable 36, 74
  Vigenère Chapter 3
cipher system 3
  strength of 7
classical sieve formula 190
Clipper 187
code 5, Chapter 6
  error-detecting 8, 164
  Hamming 8
  ISBN 8
  Italian naval 67
  Japanese naval 67
  Mengarini 67
  Morse 65
  one-part 65–6
  two-part 66–7
  U-boat 67
code-book 4
codebreaker 4
codebreaking ix, 4
code group 5, 38, 65
coin spinning 95–6
Colossus 159
combinations 207
combinatorics 206

computers
  early 161
  Manchester 161–2
  multi-access 162
  networks 162
congruent 10
continued fraction 194, 214
core store 162
cosmic ray 97
coupon collector's problem 205
crib-dragging 80
crosswords ix
cryptanalysis 4
cryptanalyst 4
cryptographer 4
cryptographic system 3
cryptography 4

decipherment 3
decryption 3
delay line 161
delta-ing 144
depth 34, 93, 111, 122, 205
  recognising 34–5
DES 143, 169
  chaining, use of 186
  encipherment/decipherment 183–4
  implementation 186
  meet-in-the-middle attack 215
  + RSA 186–7
  security of 184
  triple encipherment 185
dice 96
Diffie–Helman system 166–9
  strength of 168–9
digraph 3
  substitution 170
digraph-to-digraph cipher 58
discrete logarithm problem 168, 182
double encipherment 52, 132
double Playfair cipher 61–2
double transposition cipher 44–6

electronic mail 162
elliptic curves 189, 215–7
elliptic function (Weierstrass) 216
encipher/decipher handle on Hagelin 133
encipherment/decipherment in DES 183–4
  definition 3
  double 52, 132
  on Hagelin machine 135–6
  on SZ42 (diagram) 157
  triple in DES 185
  by wired wheels 116
encryption, definition 3

English (frequencies of letters) 18, 19
Enigma cipher machine Chapter 9, Plates 9.1–9.4 and *passim*
  Abwehr 124, 132
  'Achilles heel' 121–3
  aligning the chains 128
  'composite reflector' 124
  depths in messages 205
  encipherment procedure 123
  entry wheel 113
  ground setting 123
  identifying R1 128–9
  indicator chains 125, 205
  keyboard 112
  modifications to 130–1
  notch rings 112, 124, Plate 9. 2
  number of messages needed 127
  number of trials needed 121–2
  plugboard 121
  reflector 113, 204
  setting rings 112
  wheel 112, Plates 9. 1, 9. 2
  wheel motion 119
  wheel wirings 202–4
  Umkehrwalz *see* reflector *above*
Eratosthenes's sieve 212
error detecting/correcting code 8, 164
ETH Zürich 187
Euclidean Algorithm 177, 212–14
Euclid's proof of infinity of primes 192–3, 219
Euler's constant 205
Euler's function 211

factorial function 18
factorisation 171
Fermat–Euler Theorem 173–5, 210–11
Fermat's 'Last Theorem' 173
Fermat's 'Little Theorem' 173
Fibonacci 70
  sequence 70–1, 98, 193–4, 197–8
FORTRAN 162
French (frequencies of letters) 25

Galois field 216
'GARBO' (double agent) 9, 52, 88
GARBO's ciphers 88–92
Geiger counter 97
German (frequencies of letters) 25
German WW2 double Playfair cipher 61–3
Greeks (ancient) 2

Hagelin cipher machine Chapter 10, Plates 10.1, 10.2 and *passim*

active pin 136
cage: 'good' 141; number of 141, 206–7
decipherment 135–6
'delta-ing' process 144
encipher/decipher handle 135
encipherment 135–6
inactive pin 136
input wheel 135
key value 138
kick 135–6, 207
lug 134–6
lug cage 134–5
overlapped cage 134
overlapping 147–51
pin 134
pin wheel 134
print wheel 135
slide 147, 208; identification of 148
solving: from cipher 150–2; from key 143–7
unoverlapped 134
wheel-length 134
'work factor' 142
Hamming code 8
h.c.f., finding by Euclidean Algorithm 213–14

IBM 183
IDEA 187
identification 4
inclusion–exclusion principle 190
indicator 33, 86, 110, 122
ink (secret) 9, 72
internet Chapter 13
ISBN code 8
ITA (alphabet) 155–6
Italian (frequencies of letters) 25
Italian naval code 67

Japanese 9
kana representation 58
Japanese naval cipher (JN40) 58
Japanese naval code (OTSU) 67
Jefferson cylinder 37–9, 110, 122
'jigsaw' cipher Chapter 4

key
binary 100–3
public Chapter 12, 171, 179
transposition 40
use of two or more 42, 46, 104–6
Vigenère 29
key distribution problem 166
key value *see* Hagelin
keyword 59

kick *see* Hagelin

letter frequency 18, 24, 25, 79
in book ciphers 79, 194–5
Linear A 230
Linear B 7, 230
linear congruence generator 107–9
linear recurrence 99, 198–9, 200–2
cryptanalysis of 104
order of 99
linear sequence
as key generator 101–3
maximum period 102–3
Lorenz SZ42 cipher machine 106, Chapter 11
lottery 97
lug *see* Hagelin
lug cage *see* Hagelin

M209 cipher machine 133
Manchester University computer 161–2
Mariner spacecraft 8
matrix representation of wired wheels 118
MDTM cipher 56–8
meet-in-the-middle attack 215
Mengarini code 67
microdot 9, 72
mid-square method 106–7
modular arithmetic 10, 68, 209
monograph 3
monograph-to-digraph cipher 54
Monte Carlo method 106
Morse code 65
multi-access computing 162

Navajo Indian language 6
NBS (National Bureau of Standards) 183
notch ring *see* Enigma

octographic false 'hit' 31, 35
one-part code 65–6
one-time pad 29, 74, 92–3, 153, 195–6
one-way function 182
overlapping *see* Hagelin

parity checking 163–4
partition 206
PGP ('Pretty Good Privacy') 187
pin wheel *see* Hagelin
Playfair cipher 59–60
double 61–2
German WW2 61–2
plugboard *see* Enigma
Poisson distribution 196
Polish cryptanalysts 125–6, 128

polyalphabetic systems Chapter 3
polygraph 2
prime
  finding large 167
  'probable' 176, 212
Prime Number Theorem 209
pseudo-random number 95, 133, 202
pseudo-random number generator 106
pseudo-random sequence 95–98
public key system 171, 182

random letter 95
random number 94–5
random sequence 94
  production of 95–6
reflector *see* Enigma
repeated squaring, efficiency of 214–15
Rosetta stone 230
roulette wheel 96
RSA cipher 167–8, 175–82, 212–15
  + DES 186–7
  encipher/decipher key 175–6
  encipher/decipher process 176

sampling with replacement 205
security 163
sequence
  Fibonacci 70–1, 98, 193–4, 197–8
  linear 99–103
  pseudo-random 95–8
  random 94–5
setting, cipher machine 4
setting ring *see* Enigma
sieve, classical formula 190
signature verification 165, 188–9
simple substitution cipher Chapter 2 and *passim*
  how to solve 17–19
  minimum length for solving 16, 26
Skipjack, encryption algorithm 187
slide *see* Hagelin
spy Chapter 7
stencil cipher 73–4
strength of cipher system 7
string 3
  length of 3
symbol 3
SZ42 cipher machine 106, Chapter 11
  encipherment process 156–7
  modification 159–60
  wheel motion 155–6
  work factor 158

transistor 162
transpositions ix, 40, 55, 74, 90
  double 44–6, 89–92
  irregular box 50
  method of solution 42
  regular box 48
  security of 51
  simple 40
trapdoor 143, 185
trigraph 3
Trojan horse attack 163
two-letter cipher 54
T52 cipher machine 154

U-boat code 67
Umkehrwalze *see* Enigma (reflector)
unbreakable cipher 29, 36, 74

Vigenère cipher 28–9, 79
  how to solve 29, 30
  key 29
  text needed 37
Virus 163
Voynich Manuscript 230

Welsh (frequencies of letters) 25
wheels *see* Enigma, Hagelin, SZ42
wheel wiring *see* Enigma
wired wheel, encipherment by 116
Wittgenstein, diary of 230
World War 1 111
  1918 German cipher 57, 61
World War 2 2, 6, 58, 67, 106, 133
  double agent *see* GARBO
  German double Playfair cipher 61–3
  Italian naval cipher (Mengarini) 67
  Japanese naval cipher (JN40) 58
  Japanese naval code (OTSU) 67
  U-boat code 67

Zimmermann telegram 65–6